MOTION

TASK CARD SERIES

Conceived and
written by
Ron Marson
Illustrated by
Peg Marson

 LEARNING SYSTEMS

10970 S. Mulino Rd.
Canby OR 97013

ISBN 0-941008-98-3

Printed on Recycled Paper ♻

CONTENTS

A TOPS Model for Effective Science Teaching...

If science were only a set of explanations and a collection of facts, you could teach it with blackboard and chalk. You could assign students to read chapters and answer the questions that followed. Good students would take notes, read the text, turn in assignments, then give you all this information back again on a final exam. Science is traditionally taught in this manner. Everybody learns the same body of information at the same time. Class togetherness is preserved.

But science is more than this.

Science is also process — a dynamic interaction of rational inquiry and creative play. Scientists probe, poke, handle, observe, question, think up theories, test ideas, jump to conclusions, make mistakes, revise, synthesize, communicate, disagree and discover. Students can understand science as process only if they are free to think and act like scientists, in a classroom that recognizes and honors individual differences.

Science is *both* a traditional body of knowledge *and* an individualized process of creative inquiry. Science as process cannot ignore tradition. We stand on the shoulders of those who have gone before. If each generation reinvents the wheel, there is no time to discover the stars. Nor can traditional science continue to evolve and redefine itself without process. Science without this cutting edge of discovery is a static, dead thing.

Here is a teaching model that combines the best of both elements into one integrated whole. It is only a model. Like any scientific theory, it must give way over time to new and better ideas. We challenge you to incorporate this TOPS model into your own teaching practice. Change it and make it better so it works for you.

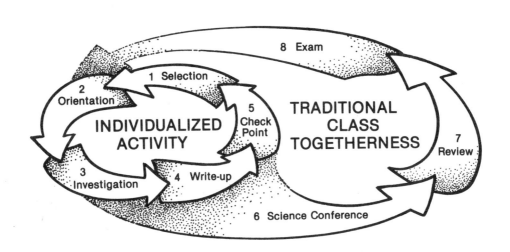

1. SELECTION

Doing TOPS is as easy as selecting the first task card and doing what it says, then the second, then the third, and so on. Working at their own pace, students fall into a natural routine that creates stability and order. They still have questions and problems, to be sure, but students know where they are and where they need to go.

Students generally select task cards in sequence because new concepts build on old ones in a specific order. There are, however, exceptions to this rule: students might *skip* a task that is not challenging; *repeat* a task with doubtful results; *add* a task of their own design to answer original "what would happen if" questions.

2. ORIENTATION

Many students will simply read a task card and immediately understand what to do. Others will require further verbal interpretation. Identify poor readers in your class. When they ask, "What does this mean?" they may be asking in reality, "Will you please read this card aloud?"

With such a diverse range of talent among students, how can you individualize activity and still hope to finish this module as a cohesive group? It's easy. By the time your most advanced students have completed all the task cards, including the enrichment series at the end, your slower students have at least completed the basic core curriculum. This core provides the common background so necessary for meaningful discussion, review and testing on a class basis.

3. INVESTIGATION

Students work through the task cards independently and cooperatively. They follow their own experimental strategies and help each other. You encourage this behavior by helping students only *after* they have tried to help themselves. As a resource person, you work to stay *out* of the center of attention, answering student questions rather than posing teacher questions.

When you need to speak to everyone at once, it is appropriate to interrupt individual task card activity and address the whole class, rather than repeat yourself over and over again. If you plan ahead, you'll find that most interruptions can fit into brief introductory remarks at the beginning of each new period.

4. WRITE-UP

Task cards ask students to explain the "how and why" of things. Write-ups are brief and to the point. Students may accelerate their pace through the task cards by writing these reports out of class.

Students may work alone or in cooperative lab groups. But each one must prepare an original write-up. These must be brought to the teacher for approval as soon as they are completed. Avoid dealing with too many write-ups near the end of the module, by enforcing this simple rule: each write-up must be approved *before* continuing on to the next task card.

5. CHECK POINT

The student and teacher evaluate each write-up together on a pass/no-pass basis. (Thus no time is wasted haggling over grades.) If the student has made reasonable effort consistent with individual ability, the write-up is checked off on a progress chart and included in the student's personal assignment folder or notebook kept on file in class.

Because the student is present when you evaluate, feedback is immediate and effective. A few seconds of this direct student-teacher interaction is surely more effective than 5 minutes worth of margin notes that students may or may not heed. Remember, you don't have to point out every error. Zero in on particulars. If reasonable effort has not been made, direct students to make specific improvements, and see you again for a follow-up check point.

A responsible lab assistant can double the amount of individual attention each student receives. If he or she is mature and respected by your students, have the assistant check the even-numbered write-ups while you check the odd ones. This will balance the work load and insure that all students receive equal treatment.

6. SCIENCE CONFERENCE

After individualized task card activity has ended, this is a time for students to come together, to discuss experimental results, to debate and draw conclusions. Slower students learn about the enrichment activities of faster students. Those who did original investigations, or made unusual discoveries, share this information with their peers, just like scientists at a real conference. This conference is open to films, newspaper articles and community speakers. It is a perfect time to consider the technological and social implications of the topic you are studying.

7. READ AND REVIEW

Does your school have an adopted science textbook? Do parts of your science syllabus still need to be covered? Now is the time to integrate other traditional science resources into your overall program. Your students already share a common background of hands-on lab work. With this shared base of experience, they can now read the text with greater understanding, think and problem-solve more successfully, communicate more effectively.

You might spend just a day on this step or an entire week. Finish with a review of key concepts in preparation for the final exam. Test questions in this module provide an excellent basis for discussion and study.

8. EXAM

Use any combination of the review/test questions, plus questions of your own, to determine how well students have mastered the concepts they've been learning. Those who finish your exam early might begin work on the first activity in the next new TOPS module.

Now that your class has completed a major TOPS learning cycle, it's time to start fresh with a brand new topic. Those who messed up and got behind don't need to stay there. Everyone begins the new topic on an equal footing. This frequent change of pace encourages your students to work hard, to enjoy what they learn, and thereby grow in scientific literacy.

GETTING READY

Here is a checklist of things to think about and preparations to make before your first lesson.

☐ Decide if this TOPS module is the best one to teach next.

TOPS modules are flexible. They can generally be scheduled in any order to meet your own class needs. Some lessons within certain modules, however, do require basic math skills or a knowledge of fundamental laboratory techniques. Review the task cards in this module now if you are not yet familiar with them. Decide whether you should teach any of these other TOPS modules first: *Measuring Length, Graphing, Metric Measure, Weighing* or *Electricity* (before *Magnetism*). It may be that your students already possess these requisite skills or that you can compensate with extra class discussion or special assistance.

☐ Number your task card masters in pencil.

The small number printed in the lower right corner of each task card shows its position within the overall series. If this ordering fits your schedule, copy each number into the blank parentheses directly above it at the top of the card. Be sure to use pencil rather than ink. You may decide to revise, upgrade or rearrange these task cards next time you teach this module. To do this, write your own better ideas on blank 4 x 6 index cards, and renumber them into the task card sequence wherever they fit best. In this manner, your curriculum will adapt and grow as you do.

☐ Copy your task card masters.

You have our permission to reproduce these task cards, for as long as you teach, with only 1 restriction: please limit the distribution of copies you make to the students you personally teach. Encourage other teachers who want to use this module to purchase their *own* copy. This supports TOPS financially, enabling us to continue publishing new TOPS modules for you. For a full list of task card options, please turn to the first task card masters numbered "cards 1-2."

☐ Collect needed materials.

Please see the opposite page.

☐ Organize a way to track completed assignment.

Keep write-ups on file in class. If you lack a vertical file, a box with a brick will serve. File folders or notebooks both make suitable assignment organizers. Students will feel a sense of accomplishment as they see their file folders grow heavy, or their notebooks fill up, with completed assignments. Easy reference and convenient review are assured, since all papers remain in one place.

Ask students to staple a sheet of numbered graph paper to the inside front cover of their file folder or notebook. Use this paper to track each student's progress through the module. Simply initial the corresponding task card number as students turn in each assignment.

☐ Review safety procedures.

Most TOPS experiments are safe even for small children. Certain lessons, however, require heat from a candle flame or Bunsen burner. Others require students to handle sharp objects like scissors, straight pins and razor blades. These task cards should not be attempted by immature students unless they are closely supervised. You might choose instead to turn these experiments into teacher demonstrations.

Unusual hazards are noted in the teaching notes and task cards where appropriate. But the curriculum cannot anticipate irresponsible behavior or negligence. It is ultimately the teacher's responsibility to see that common sense safety rules are followed at all times. Begin with these basic safety rules:

1. Eye Protection: Wear safety goggles when heating liquids or solids to high temperatures.
2. Poisons: Never taste anything unless told to do so.
3. Fire: Keep loose hair or clothing away from flames. Point test tubes which are heating away from your face and your neighbor's.
4. Glass Tubing: Don't force through stoppers. (The teacher should fit glass tubes to stoppers in advance, using a lubricant.)
5. Gas: Light the match first, before turning on the gas.

☐ Communicate your grading expectations.

Whatever your philosophy of grading, your students need to understand the standards you expect and how they will be assessed. Here is a grading scheme that counts individual effort, attitude and overall achievement. We think these 3 components deserve equal weight:

1. Pace (effort): Tally the number of check points you have initialed on the graph paper attached to each student's file folder or science notebook. Low ability students should be able to keep pace with gifted students, since write-ups are evaluated relative to individual performance standards. Students with absences or those who tend to work at a slow pace may (or may not) choose to overcome this disadvantage by assigning themselves more homework out of class.

2. Participation (attitude): This is a subjective grade assigned to reflect each student's attitude and class behavior. Active participators who work to capacity receive high marks. Inactive onlookers, who waste time in class and copy the results of others, receive low marks.

3. Exam (achievement): Task cards point toward generalizations that provide a base for hypothesizing and predicting. A final test over the entire module determines whether students understand relevant theory and can apply it in a predictive way.

Gathering Materials

Listed below is everything you'll need to teach this module. You already have many of these items. The rest are available from your supermarket, drugstore and hardware store. Laboratory supplies may be ordered through a science supply catalog. Hobby stores also carry basic science equipment.

Keep this classification key in mind as you review what's needed:

special in-a-box materials:	general on-the-shelf materials:
Italic type suggests that these materials are unusual. Keep these specialty items in a separate box. After you finish teaching this module, label the box for storage and put it away, ready to use again the next time you teach this module.	Normal type suggests that these materials are common. Keep these basics on shelves or in drawers that are readily accessible to your students. The next TOPS module you teach will likely utilize many of these same materials.
(substituted materials):	***optional materials:***
A parentheses following any item suggests a ready substitute. These alternatives may work just as well as the original, perhaps better. Don't be afraid to improvise, to make do with what you have.	An asterisk sets these items apart. They are nice to have, but you can easily live without them. They are probably not worth the extra trip, unless you are gathering other materials as well.

Everything is listed in order of first use. Start gathering at the top of this list and work down. Ask students to bring recycled items from home. The teaching notes may occasionally suggest additional student activity under the heading "Extensions." Materials for these optional experiments are listed neither here nor in the teaching notes. Read the extension itself to find out what new materials, if any, are required.

Needed quantities depend on how many students you have, how you organize them into activity groups, and how you teach. Decide which of these 3 estimates best applies to you, then adjust quantities up or down as necessary:

$Q_1 / Q_2 / Q_3$
- **Single Student:** Enough for 1 student to do all the experiments.
- **Individualized Approach:** Enough for 30 students informally working in 10 lab groups, all self-paced.
- **Traditional Approach:** Enough for 30 students, organized into 10 lab groups, all doing the same lesson.

KEY: *special in-a-box materials* general on-the-shelf materials
(substituted materials) *optional materials*

$Q_1 / Q_2 / Q_3$

1/10/10 meter sticks	3 /15/30 baby food jars
8/80/80 meters of string	1/4/10 narrow-mouth bottles (Erlynmeyer flasks)
20/200/200 *size #16 rubber bands* — see activity 13	1/1/1 roll aluminum foil
5/50/50 books to incline a meter stick (any solid support)	1/3/3 boxes paper clips - must have uniform size
1/2/2 rolls masking tape	1/1/1 corrugated cardboard box at least 30 cm high
1/10/10 paper drinking cups	1/2/5 paper punches
1/10/10 scissors	1/10/10 spring scales with a 2 or 3 N capacity, 2.5 N (250 grams) is ideal — see activity 14
3/30/30 marbles	1/4/10 *plastic gallon milk jugs with handle*
8/60/80 pennies	1/4/10 flat washers
2/20/20 sheets lined notebook paper	1/ 4/10 protractors
1/10/10 *small flat buttons*	1/10/10 manila file folders
1/1/1 roll adding-machine tape	1/5/10 plastic straws (glass tubing with smooth fire-polished ends works even better)
1/5/10 felt-tipped pens	
1/1/1 spool thread	1/1/1 *cotton ball (feathers)*
1/1/1 wall clock with second hand (watches)	1/20/30 clothespins
1/10/10 *calculators*	1/10/10 balloons
5/50/50 index cards, 4x6 or larger	1/10/10 flexible plastic drinking straws
1/1/1 bottle dishwashing liquid (bar of soap)	1/10/10 straight pins
1/1/1 bottle food coloring	1/4/10 *bath or beach towels*
1/5/10 jar lids or crucibles	1/10/10 stopwatches
1 /5/10 Ping-Pong balls	1/4/10 *pillows (coats)*
.5/5/5 cups of oil-based clay	

Sequencing Task Cards

This logic tree shows how all the task cards in this module tie together. In general, students begin at the trunk of the tree and work up through the related branches. As the diagram suggests, the way to upper level activities leads up from lower level activities.

At the teacher's discretion, certain activities can be omitted or sequences changed to meet specific class needs. The only activities that must be completed in sequence are indicated by leaves that open *vertically* into the ones above them. In these cases the lower activity is a prerequisite to the upper.

When possible, students should complete the task cards in the same sequence as numbered. If time is short, however, or certain students need to catch up, you can use the logic tree to identify concept-related *horizontal* activities. Some of these might be omitted since they serve only to reinforce learned concepts rather than introduce new ones.

On the other hand, if students complete all the activities at a certain horizontal concept level, then experience difficulty at the next higher level, you might go back down the logic tree to have students repeat specific key activities for greater reinforcement.

For whatever reason, when you wish to make sequence changes, you'll find this logic tree a valuable reference. Parentheses in the upper right corner of each task card allow you total flexibility. They are left blank so you can pencil in sequence numbers of your own choosing.

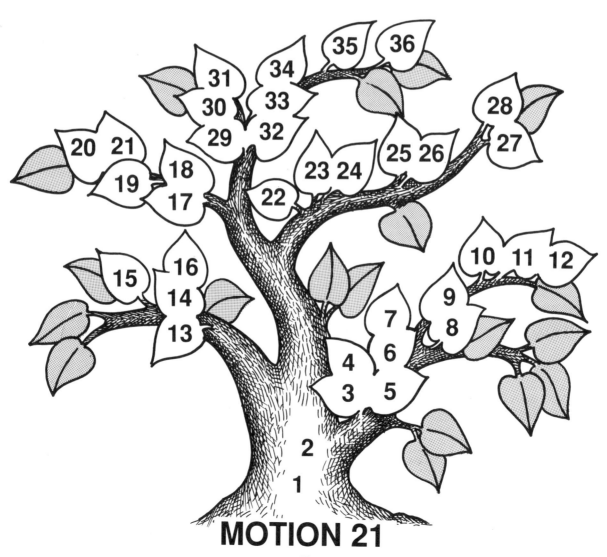

MOTION 21

E

LONG-RANGE OBJECTIVES

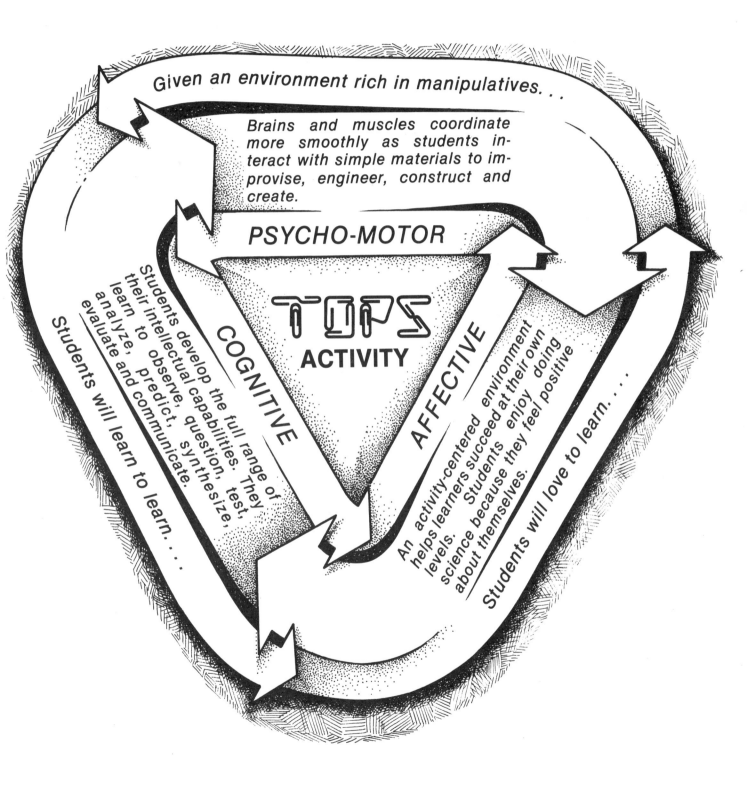

Given an environment rich in manipulatives. . .

Brains and muscles coordinate more smoothly as students interact with simple materials to improvise, engineer, construct and create.

PSYCHO-MOTOR

TOPS
ACTIVITY

COGNITIVE

Students develop the full range of their intellectual capabilities. They learn to observe, question, test, analyze, predict, synthesize, evaluate and communicate.

Students will learn to learn. . . .

AFFECTIVE

An activity-centered environment helps learners succeed at their own levels. Students enjoy doing science because they feel positive about themselves.

Students will love to learn. . . .

F

Review / Test Questions

Photocopy the questions below. On a separate sheet of blank paper, cut and paste those boxes you want to use as test questions. Include questions of your own design, as well. Crowd all these questions onto a single page for students to answer on another paper, or leave space for student responses after each question, as you wish. Duplicate a class set and your custom-made test is ready to use. Use leftover questions as a review in preparation for the final exam.

tasks 1-2
Winning football teams tend to have big men in the lineup. What advantage does extra mass give to…
a. a player at rest?
b. a player in motion?

task 2
Before going down a long hill, road signs often warn truck drivers to test their brakes. Passenger car drivers are not cautioned to do this. Why?

tasks 3-7
The position of each car is marked after each second of travel. The distance between marks is 15 meters for car (x) and 25 meters for car (y).

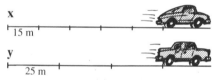

a. How fast does each car move?
b. Draw a data table and graph. Label each graph line.

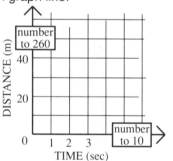

c. Interpret your graph: Why are the graph lines straight? Why does one have a steeper slope?

tasks 5-7
You drive about 990 km from Denver to Kansas City in 11 hours, stopping exactly 1 hour for lunch in Salina along the way.
a. What is your average speed?
b. What is your estimated freeway speed.?
c. Why is your average speed different than your freeway speed?
d. If you stopped for lunch after 7 hours, what is the distance between Denver and Salina?

tasks 8-9
Consider these separate collision events labeled x, y, and z. Assume that all balls move at the same speed.

a. Which are elastic; inelastic? Why?
b. Which involve balls of unequal mass? Explain.
c. Which involve balls of equal mass? Explain.

tasks 10-12
In a two-car accident, one driver suffers whiplash (a neck injury from the head being tossed back). The other driver suffers a broken nose. Use the idea of inertia to explain how this accident likely happened.

tasks 10-12
Explain each statement in terms of Newton's first law of motion.
a. The head of a hammer is made from heavy iron.
b. You pitch a shovelful of dirt; when the shovel stops, the dirt flies off.

task 13
A chain is pulled with increasing force until it breaks. Can you predict which link will fail first?

task 14
This simple scale is calibrated in Newtons. Describe how its rubber band responds to increasing amounts of force.

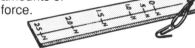

task 15
Is more force required to start a book sliding, or keep a book sliding at uniform speed? Explain.

tasks 15,18
Use labeled arrows to illustrate *all* forces that act upon each brick. Say if they are balanced or unbalanced.
a. A brick rests on the table.
b. A brick slows down as it slides across the floor.
c. You push lightly on a brick, but it doesn't move.
d. A brick drops to the floor. It hasn't yet landed.

tasks 15, 18
Tell if the forces acting on each object are balanced or unbalanced. Explain how you know.
a. A car moves down a flat road at a constant 50 km/hr.
b. A snowflake falls to earth on a silent, windless day.
c. The moon orbits the earth.
d. This test paper rests motionless on the table.
e. A stone is tossed straight up into the air.

task 16
A gymnast weights 500 N. How much tension is on each arm when she hangs from an overhead bar by…
a. 1 arm?
b. 2 parallel arms?
c. 2 widely separated arms? Explain.

task 16
Two adults and a child pull as hard as they can on a 3-way rope. Both adults have equal strength; the child half as much as either one. If no one is able to move the other, show the relative magnitudes and directions of each applied force. (Draw a bird's-eye perspective. Label your arrows.)

tasks 17-18
You are standing on a bus facing forward. Your legs are relaxed because you are firmly grasping an overhead bar. What is the bus doing if you find yourself…
a. leaning back?
b. leaning right?
c. leaning forward?
d. not leaning at all?

task 18
Light travels in a straight line at a constant speed of 300,000 km/s. What is its acceleration?

task 19
A stone falls 10 m/s after 1 second, 20 m/s after 2 seconds, 30 m/s after 3 seconds, and so on.
a. What is its acceleration?
b. If you plotted time vs. distance, would you graph a straight line?

Answers

tasks 1-2

a. A player at rest has greater mass, and therefore a greater tendency to remain at rest. It takes more force for other players to push him out of the way, or more time to go around.

b. The moving player has greater mass, and therefore a greater tendency to remain in motion. This enables him to push other players out of the way more easily.

task 2

Trucks have far greater mass than passenger cars. As a consequence they have a much greater tendency to stay in motion, requiring more breaking force to slow down or stop.

tasks 3-7

a. Car (x) moves 15 m/sec. Car (y) moves 25 m/sec.

b.

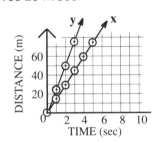

c. The graph lines are straight because each car moves at a uniform speed. The graph line for car (y) has a steeper slope because it moves at a faster constant speed than car (x).

tasks 5-7

a. average speed = 990 km/11 hr
= 90 km/hr

b. freeway speed = 990 km/10 hr
= 99 km/hr

c. Your average speed is slower because it is based on the total time of your journey, including the time you stopped for lunch.

d. 99 km/hr x 7 hr = 693 km

tasks 8-9

a. Collisions x and y are elastic because the balls rebound without sticking together. Collision z is inelastic because the balls stick together.

b. Collision y involves balls of unequal mass. Since the left one is deflected less than the right, it has greater mass.

c. Collisions x and z involve balls of equal mass. The balls are deflected equally in collision x, and both come to a complete stop in collision z.

tasks 10-12

The driver with the broken nose likely "tail-ended" the driver with whiplash. As the moving car came to a crashing halt, the driver's head, due to inertia, continued to move forward, breaking his nose against the dashboard. Meanwhile, the driver in the stationary (or slower moving) car suddenly lurched forward in response to the impact from behind. The drivers body accelerated forward but her unsupported head, due to inertia, remained behind.

tasks 10-12

a. Because of its extra mass or inertia, an iron hammer head will drive the nail deeper into the wood before stopping.

b. You hang onto the shovel after pitching the dirt. This slows and stops the shovel, but the dirt, because of inertia, continues moving forward in the same direction.

task 13

No. Tension is the same throughout the chain. Since all links are exposed to equal force, any one could break first.

task 14

It stretches more with each Newton of added force. At first the change in length per Newton is less; later on, it stretches in equal increments, in direct proportion to the applied force.

task 15

More force is required to start the book sliding because both its inertia and static friction must be overcome. To keep the book sliding at uniform speed, however, only its moving friction needs to be balanced. This moving friction is less than static friction because irregularities in the surface of the book and table cannot lock together as long as the moving surfaces continuously slide by each other.

tasks 15,18

a. Balanced.

b. Unbalanced.

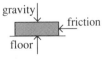

c. Balanced.   d. Unbalanced.

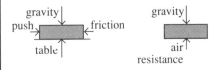

tasks 15, 18

a. Balanced forces: the car is moving at a constant speed; it does not accelerate.

b. Balanced forces: the snowflake drifts down at a slow, constant speed; there is no acceleration.

c. Unbalanced forces: The moon is continuously changing directions (accelerating) as it orbits earth.

d. Balanced forces: The test paper is at rest; it does not accelerate.

e. Unbalanced forces: The rock slows down as it climbs (decelerates), and speeds up as it falls back down (accelerates).

task 16

a. 500 N.   b. 250 N.

c. > 250 N. More force is required on each arm since only its vertical component balances the downward pull of gravity.

task 16

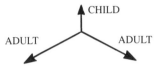

tasks 17-18

The bus is...

a. accelerating.

b. turning (accelerating) left

c. decelerating.

d. moving at a constant speed or at rest.

task 18

It has no acceleration because the speed of light is constant.

task 19

a. a = 10 m/s/s or 10 m/s^2.

b. No. Constant motion graphs as a straight line. In this case the graph line would sweep up into a curve.

task 20

task 21

The earth's orbit is a compromise between its inertial tendency to move in a straight line and its acceleration toward the center of the sun by the unbalanced force of gravity.

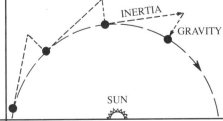

H

Review / Test Questions (continued)

task 20

At a football game you watch the quarterback throw a "desperation pass" as far as possible down the field.

a. Sketch the flight path of the football. Use an arrow to show the direction it moves.

b. Label where the ball accelerates and decelerates along its path.

task 21

Gravity constantly attracts the earth and sun together. Yet the earth remains in orbit around the sun without crashing into it. Explain how this is possible. Use a diagram to illustrate your answer.

task 22

Explain each event in terms of Newton's second law;

a. A passenger car accelerates away from a traffic light better than a heavy truck.

b. Adults usually throw baseballs farther than children.

task 23

A 1 kg mass weighs about 10 Newtons. Use Newton's second law to calculate the acceleration of gravity.

task 24

A feather and a marble are sealed in a large jar and dropped from a high bridge. Where does each object rest in the jar on the way down? Explain.

task 25

A little girl and her father are both standing on roller skates.

a. What happens to the little girl as she pushes her father forward? Frame your answer in terms of Newton's third law.

b. Who travels farthest? Use Newton's second law to support your answer.

task 26

It is easier to start running on a cinder track than on a sheet of ice. Use Newton's third law to explain why.

task 26

You are in a small boat with no oars. But you do have a load of heavy stones. Use Newton's third law to propose a way of propelling the boat forward with these stones.

task 27-28

Design a sling shot and projectile to travel a maximum distance.

task 29-30

A piano falls out the window of a very tall building. The distance (d) through which it falls is given by $d = 1/2gt^2$, where (g) is the acceleration of gravity and (t) is the time in seconds. Because g is nearly 10 m/s^2 on earth, the distance in meters for this equation is given by $d = 5 t^2$.

a. Complete this table of time vs. distance over the first 125 meters.

b. Graph your results.

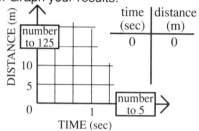

time (sec)	distance (m)
0	0

c. Interpret the physical significance of the graph line.

task 31

A piano falls out the window of a very tall building. (It is the same piano. Throwing out a new one would be too expensive.)

a. Knowing the acceleration of gravity on earth to be about 10 m/s^2, complete this table of time vs. speed over the first 5 seconds.

b. Graph your results.

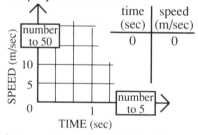

time (sec)	speed (m/sec)
0	0

c. Interpret the physical significance of the graph line.

task 32

Explain how the force of gravity (a marble's weight) is resolved by an inclined plane into a smaller accelerating force. Illustrate your answer with a labeled diagram.

task 33

A meter stick is tipped at just the right angle so that a marble accelerates through 20 cm in 1 second.

a. How far would the same marble travel down the same incline in 2 seconds?

b. In what fraction of a second would a marble travel 5 cm down this incline?

task 34

A marble clicks out 7 equally-spaced units of time when it rolls across strings positioned at these intervals on an inclined meter stick.

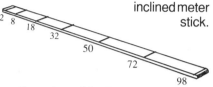

a. Complete this data over the first 7 clicks for the distance shown.

b. Graph time^2 vs, distance.

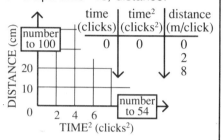

time (clicks)	time^2 (clicks^2)	distance (m/click)
0	0	0
		2
		8

c. Interpret your graph line.

tasks 10-12, 24, 35

A bicycle and a fully loaded cement truck are at rest, balancing on the top of a long steep hill ready to begin a "gravity race." At the command "go" the cyclist and truck driver both push their respective vehicles as hard as they can, then jump aboard to coast to the finish line several miles away. Use your knowledge of Newton's laws of motion to predict…

a. How the race will start.

b. How the race will finish.

c. How a similar race on the moon would be different.

task 36

It is difficult to catch a dollar bill that a friend drops through your fingers if you start at the center and don't lower your hand.

a. Would this trick be easier to do on the moon? (You are not wearing a space suit because you are inside a protective bubble.)

b. Where on earth could you make this trick easier to do?

I

Answers (continued)

task 22

a. Newton's second law says that acceleration is inversely proportional to mass ($a = F/m$). As mass increases, therefore, from the lighter passenger car to the heavier truck, acceleration is dramatically reduced. Thus the car easily speeds ahead of the truck.

b. Newton's second law says that acceleration is directly proportional to force ($a = F/m$). Because adults throws baseballs with more force than children, they accelerate them over longer distances.

task 23

By Newton's second law, $a = F/m$. Thus the acceleration of gravity is given by...

$$a = \frac{10\ N}{1\ kg} = \frac{10\ m}{sec^2}$$

task 24

If there were no air resistance, both the marble and feather would float in the jar since gravity accelerates all 3 objects equally. However, the jar is slowed somewhat by air resistance. The jar, in turn, slows the marble and feather, causing them to rest gently at the lowest end of the jar.

task 25

a. As the girl exerts a force upon her father, Newton's 3rd law predicts that the father exerts an equal and opposite force upon the little girl. Both, therefore, accelerate in opposite directions.

b. Newton's second law states that acceleration is inversely proportional to mass. The little girl's smaller mass will thus accelerate backward more than her father's larger mass accelerates forward.

task 26

Friction enables you to exert force against the cinder track. It pushes back with an equal and opposite force, accelerating you forward. On ice, by contrast, lack of friction doesn't allow you to exert enough action force to produce the needed reaction force.

task 26

Throw the stones off the back of the boat. The equal and opposite reaction to each throw will push you forward, in accordance with Newton's 3rd law.

tasks 27-28

The slingshot should apply maximum force to the projectile. Thus, the rubber band should be just thick enough to allow you to draw it back the full length of your arm using your maximum strength. The projectile should have a mass that is small enough to allow for its rapid acceleration, but not so light that its inertia is easily overcome by air resistance. A steel marble might work well. Its round smooth shape would cut air resistance to a minimum.

tasks 29-30

a-b.

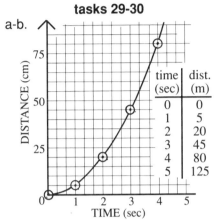

time (sec)	dist. (m)
0	0
1	5
2	20
3	45
4	80
5	125

c. The graph line curves up, showing that the piano accelerates — that distance per second (speed) increases over each new second.

task 31

a-b.

time (sec)	distance (m)
0	0
1	10
2	20
3	30
4	40
5	50

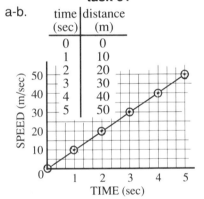

c. The straight graph line shows that the speed of the piano increases the same amount each second. In other words, acceleration is constant.

task 32

Gravity is resolved into 2 smaller components that run perpendicular and parallel to the incline. The perpendicular force is balanced by the incline pushing up in an equal and opposite direction. Only the relatively smaller parallel vector accelerates the marble down the incline.

WEIGHT

task 33

The distance (d) a marble rolls down an incline, and the square of the time it takes (t^2), are directly proportional. Thus...

a. $$\frac{20\ cm}{(1\ sec)^2} = \frac{d}{(2\ sec)^2}$$
$$d = 80\ cm$$

b. $$\frac{20\ cm}{(1\ sec)^2} = \frac{5\ cm}{(t\ sec)^2}$$
$$t = .5\ sec$$

task 34

a-b.

t	t^2	dist.
0	0	0
1	1	2
2	4	8
3	9	18
4	16	32
5	25	50
6	36	72
7	49	98

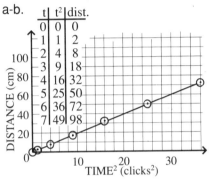

c. The straight graph line shows that the distance a marble rolls down an incline is directly proportional to the square of the time it takes.

tasks 10-12, 24, 35

a. The truck driver will need to apply a great deal of force to overcome the cement truck's huge inertia, and slowly accelerate it over the edge of the hill. Meanwhile the cyclist will quickly accelerate the light bicycle into an early lead.

b. As both vehicles gather speed, air resistance plays an important role. The passing wind will balance the force of gravity on the light bicycle much sooner than on the heavy truck. The truck will gain on the bicycle, and win the race if the hill is long enough.

c. Both vehicles have the same mass everywhere in the universe. So a race on the moon would start the same. With less gravity, both would accelerate down a similar incline more slowly. Without air resistance the bicycle would retain its early lead and the truck would never catch up. Gravity accelerates all objects, light and heavy, by the same amount.

task 36

a. Yes. On the moon, the bill would accelerate more slowly, making it easier to catch.

b. In a rapidly descending elevator. (Or in a large barrel plunging over Niagara Falls.)

TEACHING NOTES
For Activities 1-36

Task Objective (TO) experience how the total mass of a resting body affects its tendency to remain at rest.

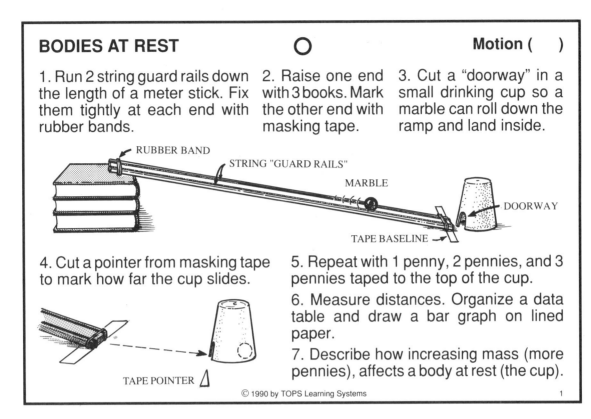

BODIES AT REST ○ Motion ( )

1. Run 2 string guard rails down the length of a meter stick. Fix them tightly at each end with rubber bands.

2. Raise one end with 3 books. Mark the other end with masking tape.

3. Cut a "doorway" in a small drinking cup so a marble can roll down the ramp and land inside.

RUBBER BAND
STRING "GUARD RAILS"
MARBLE
DOORWAY
TAPE BASELINE

4. Cut a pointer from masking tape to mark how far the cup slides.

5. Repeat with 1 penny, 2 pennies, and 3 pennies taped to the top of the cup.

6. Measure distances. Organize a data table and draw a bar graph on lined paper.

7. Describe how increasing mass (more pennies), affects a body at rest (the cup).

TAPE POINTER △

© 1990 by TOPS Learning Systems   1

Answers / Notes.

4. *The cup should slide over a reasonably large distance without being bowled over. Adjust the slope of the incline up or down as necessary by adding or removing books. Keep the tape pointers well back from the sliding area. Otherwise they can snag the cup as it slides by.*

6. *Measure distances without removing the string guard rails. These meter sticks will be used again and again as inclined planes.*

   Data will vary widely, affected by a host of variables — mass of the marble, slope of the incline, smoothness of the table surface, how the marble strikes the cup, etc. But overall trends in the data remain definite.

7. As the mass of the cup increases (with the addition of more pennies) its tendency to remain at rest increases (it moves through shorter distances).

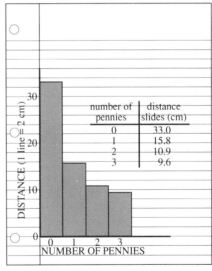

number of pennies	distance slides (cm)
0	33.0
1	15.8
2	10.9
3	9.6

DISTANCE (1 line = 2 cm)
NUMBER OF PENNIES

Materials

☐ A meter stick and string.
☐ Rubber bands.
☐ Books or other objects to raise one end of the incline.
☐ Masking tape.
☐ A disposable drinking cup. Styrofoam cups are least desirable. They are easily knocked over when captured by the marble, and tend to build up a static charge with the table top.
☐ Scissors.
☐ A marble.
☐ Pennies.
☐ Lined notebook paper.

(TO) experience how the total mass of a moving body affects its tendency to remain in motion.

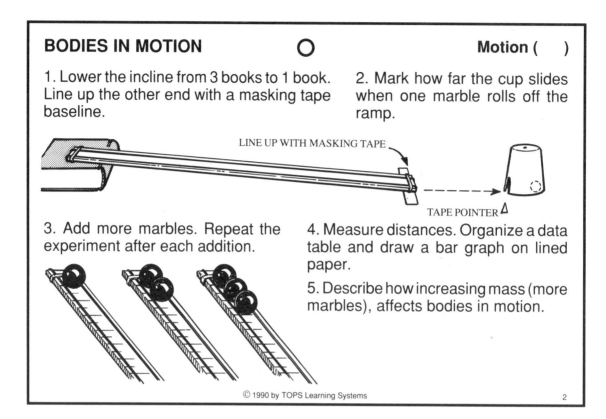

BODIES IN MOTION ○ **Motion ( )**

1. Lower the incline from 3 books to 1 book. Line up the other end with a masking tape baseline.

2. Mark how far the cup slides when one marble rolls off the ramp.

LINE UP WITH MASKING TAPE

TAPE POINTER

3. Add more marbles. Repeat the experiment after each addition.

4. Measure distances. Organize a data table and draw a bar graph on lined paper.

5. Describe how increasing mass (more marbles), affects bodies in motion.

2

Answers / Notes

1-3. Multiple marbles should roll down the incline together. Otherwise the cup accelerates away from the bottom of the ramp before all the marbles roll inside. In general, marbles that start together end together. Line them against your hand at the top of the incline, than release the lead marble by quickly removing your hand. If the marbles still won't begin rolling together, increase the slope of the incline with another book.

4. *Data will again vary widely. Here is our result:*

5. As the mass of the moving marbles increases (with the addition of more marbles to the row) their tendency to remain in motion increases (they slide the inverted cup through greater distances).

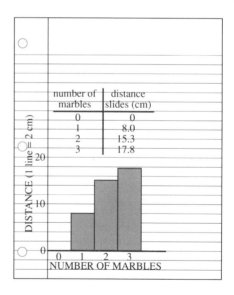

number of marbles	distance slides (cm)
0	0
1	8.0
2	15.3
3	17.8

DISTANCE (1 line = 2 cm)

NUMBER OF MARBLES

Materials

☐ Materials from activity 1.
☐ Three marbles with uniform size and weight.

(TO) mark the progress of an object sliding down an incline. To relate the spacing between these marks to speed and uniform motion.

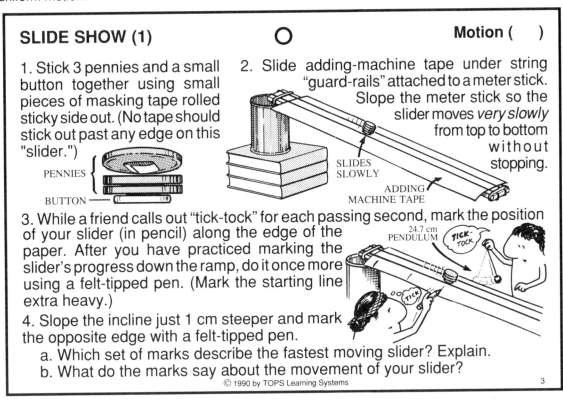

SLIDE SHOW (1)   O   **Motion ( )**

1. Stick 3 pennies and a small button together using small pieces of masking tape rolled sticky side out. (No tape should stick out past any edge on this "slider.")

PENNIES {
BUTTON —

2. Slide adding-machine tape under string "guard-rails" attached to a meter stick. Slope the meter stick so the slider moves *very slowly* from top to bottom w i t h o u t stopping.

SLIDES SLOWLY
ADDING MACHINE TAPE

3. While a friend calls out "tick-tock" for each passing second, mark the position of your slider (in pencil) along the edge of the paper. After you have practiced marking the slider's progress down the ramp, do it once more using a felt-tipped pen. (Mark the starting line extra heavy.)

24.7 cm PENDULUM   TICK-TOCK

4. Slope the incline just 1 cm steeper and mark the opposite edge with a felt-tipped pen.
    a. Which set of marks describe the fastest moving slider? Explain.
    b. What do the marks say about the movement of your slider?

© 1990 by TOPS Learning Systems   3

Answers / Notes

2-4. The slider will not move at all until the incline reaches a certain critical slope; adjust the ramp just 1 cm higher, and the slider will creep along at a remarkably slow, uniform speed. (The motion may be somewhat halting and jerky, but it still graphs into a reasonably straight line.) Move the ramp 2 cm above its critical slope and the slider moves much faster. Move the ramp higher than this and the slider begins to accelerate.

*   Because this activity is designed to study uniform motion; since data points are few and far between as the slider speeds up, the watchword for this experiment is GO SLOW! Tilt the ramp just high enough to prevent the slider from stalling midway through its run. Fine slope adjustments can be made by shifting the meter stick to ride slightly forward or backward on its support.*

4. Reverse the ramp to access the other side of the adding-machine tape. Or move to the other side of the table.

4a. The most widely spaced marks track the fastest moving slider. It traveled a relatively greater distance with each passing second.

4b. Evenly spaced marks suggest that the slider travels with uniform speed. Intervals that are somewhat shorter indicate a temporary slow-down; longer intervals, a temporary speed-up.

Materials

☐ Pennies.

☐ A flat button with a diameter somewhat smaller than a penny. If you don't have buttons handy, slide the stack of pennies directly on the meter stick. (You'll need to widen the parallel string tracks to accomodate the wider pennies.)

☐ Masking tape.

☐ A meter stick with string "guard rails".

☐ Books, cans or equivalent to raise the incline.

☐ Adding-machine tape.

☐ A felt-tipped pen.

☐ Thread to make a 24.7 cm (one-second) penny pendulum. These pendulums are unnecessary if your students have access to a room clock or wrist watches with a second hand that beats in second intervals. A continuous second sweep is not suitable.

(TO) graph the distance traveled by a sliding object as a function of time. To interpret the straightness and slope of the graph line as an indication of uniform motion and speed.

SLIDE SHOW (2)　　　　○　　　　Motion (　)

1. Measure the distance from your starting mark to each succeeding mark. Do this for both the slow and fast tracks.

2. Make a data table and plot your results. Use thread to help you determine the best possible straight line to draw among your circled points.

time (sec)	total distance (cm) slow track	fast track
0	0	0
1		
⋮	⋮	⋮

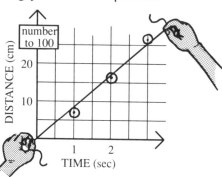

3. How is the slope of each graph line related to the speed of your slider?

4. Why don't all points touch the graph line?

4

Answers / Notes

2. Thread is a useful tool when determining the best straight line to draw through data points that are somewhat scattered. Unlike a ruler, it doesn't obscure any of the points.

Typical results:

time (sec)	total distance (cm) slow track	fast track
0	0	0
1	7.2	11.9
2	16.7	29.0
3	26.7	58.9
4	33.4	85.4
5	38.8	
6	45.0	
7	55.9	
8	61.4	
9	68.6	
10	76.3	
11	83.2	
12	89.5	

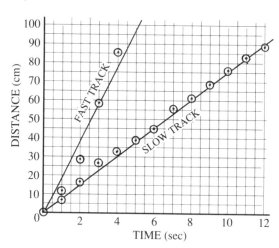

3. A steeper slope indicates a faster slider, as a shallower slope indicates a slower slider.

4. The slider did not move with uniform speed down the slope. Intervals that were slower than average had data points that dipped below the line. Faster intervals produced data points somewhat above the line.

Materials

☐ Materials from the previous activity.
☐ Thread.
☐ Graph paper. Photocopy the grid at the back of this book. (Your class will require 8-12 sheets per student, if they finish all 36 activities.)

(TO) participate in a timed race and compute average speed. To distinguish between average speed and constant speed.

HEEL-TOE SHUFFLE 〇   Motion (   )

1. Cut a piece of string 5 meters long and lay it on the floor…

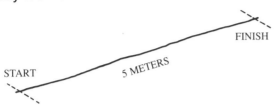

…Time how fast you can travel the length of this string, keeping *both* feet on the floor at *all* times.

2. Construct a data table. Calculate your average speed over 3 trials to the nearest .01 m/sec.

trial	distance (m)	time to shuffle 5 meters (sec)	average speed (m/sec)
1	5		
2	5		
3	5		

$$\text{AVERAGE SPEED} = \frac{\text{TOTAL DISTANCE}}{\text{TOTAL TIME}}$$

3. Does your nose move at a *constant* speed when you heel-toe shuffle? Explain the difference between *average* speed and *constant* speed.

5

Answers / Notes

2. *Shuffle times will vary widely.*

trial	distance (m)	time to shuffle 5 meters (sec)	average speed (m/sec)
1	5	7.0	.71
2	5	7.5	.67
3	5	6.0	.83

3. No. The heel-toe shuffle does not produce smooth motion. Toe down, you move faster than your average speed; heel down, slower. High, medium and low speeds all average together when you divide total distance by total time. Constant speed, by contrast, describes a specific, steady, uniform rate.

Materials

☐ A 5 meter run free of furniture. If you don't have enough space to accommodate each lab group, gather data as a single class activity, or go outside.
☐ A meter stick.
☐ Scissors.
☐ String.
☐ A watch or wall clock. If stopwatches are used, supply hand calculators as well. Long division becomes tedious when time is measured to the nearest .01 seconds.

(TO) practice moving at a nearly-constant speed. To calculate distance traveled, knowing speed and time.

HEEL-TOE WALK ○ Motion ( )

1. Practice walking at a constant speed: touch heel to toe while counting in your head at a calm, steady rate.

1001, 1002, 1003...

UNIFORM WALK

2. Ask a friend to time how long it takes you to walk the same 5-meter string course you used before. Repeat until you get consistent results. Report this as your natural speed.

trial	distance (m)	time to walk 5 meters (sec)	speed (m/sec)
1	5		
2	5		
3	5		

natural speed = ?

3. Does your nose move at a constant speed when you heel-toe walk? Compare this motion to your previous heel-toe shuffle.

4. Using your natural heel-toe walking speed, calculate how far you will travel in exactly 1 minute.

5. Test your prediction in a hallway or outside. Measure distance with your 5 meter string and a meter stick.

6. Evaluate your prediction.

© 1990 by TOPS Learning Systems

6

Answers / Notes

1. *Unlike the heel-toe shuffle, this heel-toe walk is not a race. Make it clear to your class that consistency, not speed, is their objective.*

2. *Students should repeat the 5-meter course until they get times that are internally consistent to within one second.*

trial	distance (m)	time to walk 5 meters (sec)	speed (m/sec)
1	5	24	.21
2	5	25	.20
3	5	24	.21

natural speed = .21 m/sec

3. Essentially. The motion is steady and smooth compared to the jerky heel-toe shuffle.

4. $\dfrac{.21 \text{ meters}}{\text{second}} \times 60 \text{ seconds} = 12.6 \text{ meters}$

5. *Students should measure their actual distance traveled in 60 seconds and report the result in meters.*

6. *Accept any method of comparison. Students might report a difference between the predicted distance and actual distance, or calculate a percent error.*

Materials

☐ The same 5-meter string course used in the previous activity.
☐ A wall clock, plus wrist watch or stop watch to take out of the classroom.
☐ A meter stick.

(TO) graph distance as a function of time. To interpret the shape of each graph line.

HEEL-TOE GRAPH   ◯   Motion (   )

1. Complete this data table based on your natural heel-toe walking speed from the previous activity.

2. Graph your results. Label the line "uniform walk."

time (sec)	distance (m)
0	0
10	
20	
30	
60	

3. Calculate how far you would travel in 20 seconds if your moved at your fastest average shuffle in activity 5. Plot this value on your graph. Draw a straight dashed graph-line through this point and (0,0), labeling it "fastest average shuffle."

4. Recall that you don't always shuffle at this fastest average speed, that you would soon grow tired. Draw and label a third graph line to show your "predicted progress".

5. What does a graph line's shape and steepness tell you about motion?

© 1990 by TOPS Learning Systems   7

Answers / Notes

1.

time (sec)	distance (m)
0	0
10	2.1
20	4.2
30	6.3
60	12.6

2-4.

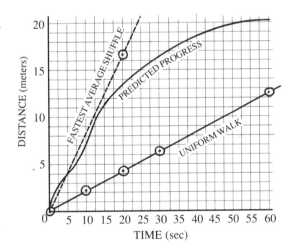

3. Our fastest average shuffle from activity 5 was .83 m/sec.

$$20 \text{ seconds} \times \frac{.83 \text{ meters}}{1 \text{ second}} = 16.6 \text{ meters}$$

5. A straight graph line of time *vs* distance implies uniform motion or constant speed: steeper slopes represent faster speeds; shallower slopes represent slower speeds. A curved graph line means nonuniform motion: as it curves up, speed increases; as it curves down, speed decreases.

Materials

☐ Graph paper.

(TO) track elastic collisions between spheres of equal mass.

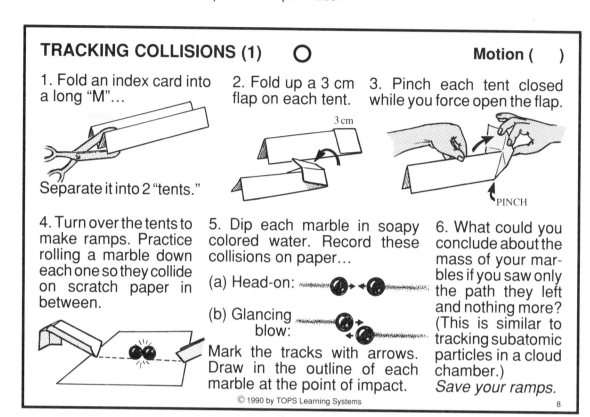

Answers / Notes

4. *You may stabilize the ramp, if necessary, by rolling bits of masking tape sticky-side-out to place underneath.*

5.　　(a) Head-on collision:　　　　　　　　　(b) Glancing blow:

6. The marbles seem to deflect each other by equal amounts, so they must have about the same size. *(More precisely, they have the same momentum, or if they have the same speed on impact, they exhibit comparable inertia or equal mass. Many of these important ideas will be developed in later activities.)*

Materials

☐ A 4 x 6 index card, or larger.
☐ Scissors.
☐ Marbles of uniform size and shape.
☐ Soapy water tinted with food coloring in a labeled jar. Add enough food coloring to tap water so it leaves a visible stain on white paper. Then add a few drops of detergent to reduce its surface tension. The solution will thereby cling to the marble and deposit a continuous track as the marble rolls across paper.
☐ Containers to hold the tracking solution. Jar lids or crucibles are convenient to use because marbles can easily be dipped in and out of the soapy tinted water without falling out of reach.
☐ A smooth level surface.
☐ Scratch paper.

(TO) track elastic collisions between spheres of unequal mass. To distinguish between elastic and inelastic collisions.

TRACKING COLLISIONS (2)   Motion (   )

1. Record these collisions between a marble and a ping pong ball, using colored soapy water on paper :
   (a) Head on;
   (b) Ping Pong ball strikes resting marble;
   (c) Marble strikes resting Ping Pong ball.

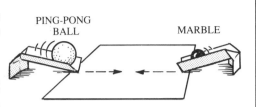

2. Mark the tracks with arrows. Draw in the outline of each sphere at the point of impact.

3. Which ball most resisted changing…
   a. Its state of motion? Explain.
   b. Its state of rest? Explain.

4. Roll some clay between your palms until it is the size and shape of a glass marble. Drop both onto your table top.
   a. Describe the *elastic* collision between the *glass* marble and the table.
   b. Describe the *inelastic* collision between the *clay* marble and the table.
   (Save your ramps.)

9

Introduction

Summarize Newton's first law of inertia on your blackboard.

> **NEWTON'S FIRST LAW (version 1)**
> Mass has *inertia*. This means…
> • a body at rest tends to stay at rest.
> • a body in motion tends to stay in motion in a straight line.

Ask why a bowling ball at rest is harder to kick than a football at rest. (It has more mass and therefore more inertia.) Then ask why a bowling ball in motion is harder to deflect or stop than a football in motion. (It has more mass and therefore more inertia.)

Answers / Notes

1-2.	a. Head-on:	b. Ping-Pong ball strikes resting marble:	c. Marble strikes resting Ping-Pong ball:

3a. The marble resisted changing its state of motion the most because it deflected the least when colliding with the Ping Pong ball. This means that the marble has more inertia (or mass) than the Ping-Pong ball.

3b. In a similar manner, the marble most resisted moving once it was in a state of rest. Compared to the Ping Pong ball at rest, it hardly moved at all. Again, this indicates that the marble has more inertia (or mass) than the Ping-Pong ball.

4a. Elastic collision: the glass marble fully rebounds and doesn't lose its spherical shape.

4b. Inelastic collision: the clay marble sticks to the table and loses its spherical shape.

Materials

☐ Use the same items as in the previous activity plus a Ping-Pong ball and clay.

(TO) experience how inertia keeps resting bodies at rest.

OUT FROM UNDER ⭕ Motion (　)

1. Trace around the mouth of a baby food jar on an index card. Cut out the circle and tape its edge to some thread about as long as notebook paper.

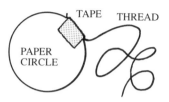

2. In each case, remove the circle, but leave the penny in place:

(a) EASY:
penny on a clothespin.

(b) HARDER:
penny on your finger.

(c) HARDEST:
penny on your finger; no thread on the circle.

3. Explain these tricks in terms of Newton's first law of inertia.

10

Answers / Notes

2. *In part a and b, if the paper is pulled away with a rapid, wrist-snapping jerk, the penny above will remain in place. Part c is more difficult because the paper must be snapped out from under the penny in a perfectly horizontal direction.*

3. Because it has mass and therefore inertia, a penny at rest tends to say at rest. As the index card circle is snapped away from beneath it, the resting penny above remains unmoved.

Materials

☐ A baby food jar.
☐ An index card.
☐ Scissors
☐ Masking tape.
☐ Thread.
☐ A clothespin. It should have a broad, flat "nose" that easily supports the penny in trick 2a. Level this surface, if necessary, with another penny placed directly under the first (underneath the paper circle). Stick it to the wood with masking tape rolled sticky-side-out.

(TO) recognize that the slow, steady application of force overcomes the inertia of a stationary book more readily than a sudden impulse does.

BOOK DROP? ○   Motion ( )

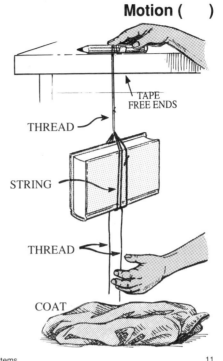

1. Wrap string several times around a heavy book and tie it. Loop 2 pieces of thread through this string, one on each side.

2. Wrap the free ends of one thread securely around your pencil. Tape the ends.

3. Suspend the book by holding the pencil along your table edge. Put a coat or something soft underneath.

4. Pull the bottom loop as directed below. Explain what happens in words and pictures:

    a. *Rapidly* pull the bottom loop with a hard, fast jerk. Does the book drop?
    b. Repeat the experiment. This time *slowly* pull the bottom loop. Does the book drop?

11

Introduction

A force may be represented by an arrow because it is a vector quantity. The length of the arrow shows its *magnitude;* the arrow head indicates the *direction* in which it is applied.

Diagram a tug-of-war on your blackboard between team A and team B. Discuss which team is stronger (B), and which team wins the contest (A).

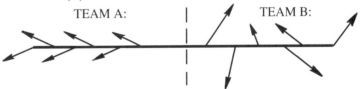

TEAM A: TEAM B:

Answers / Notes

4a. No. The mass (or inertia) of the large resting book resists moving down when the bottom thread is suddenly pulled. The bottom loop breaks under this large sudden force before it transfers the force through the book to the top loop.

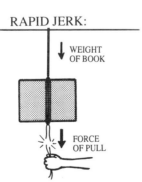

RAPID JERK:

WEIGHT OF BOOK

FORCE OF PULL

4b. Yes. The top loop breaks first because it is strained by *both* the slow pull *and* the weight of the book. The bottom loop is stressed only by the slow pull, nothing more.

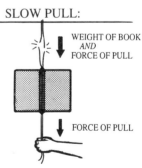

SLOW PULL:

WEIGHT OF BOOK *AND* FORCE OF PULL

FORCE OF PULL

Materials

☐ String.
☐ A heavy book. Position a blanket, coat or pillow under the book as necessary, to protect it from damage.
☐ Scissors.
☐ Thread. Light thread should be looped as illustrated to support a heavy book. If you are using heavy-duty thread, a single strand above and below the book might work better.
☐ Masking tape.
☐ A coat or pillow (optional).

(TO) apply the concept of inertia to learning a trick. To have fun.

THE HOOP TRICK ⭕   Motion ( )

1. Cut off 4 strips of index card exactly 3 spaces wide. Overlap them (in the same direction) about the width of your little finger, and secure with tape.

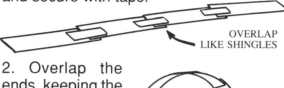

OVERLAP
LIKE SHINGLES

2. Overlap the ends, keeping the tape to the outside of the circle. Tape again to form a closed hoop.

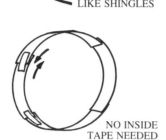

NO INSIDE
TAPE NEEDED

3. Balance the hoop over a bottle or Erlynmeyer flask. Crunch together a small wad of aluminum foil to balance on top. (It should easily fit through the mouth of the bottle.)

4. Drop the foil into the bottle by touching only the hoop with only *one* finger. Learn this trick, then challenge your friends.

5. Explain why this trick works. Illustrate your answer with a diagram.

12

Answers / Notes

4. *If students fail to figure this trick out on their own, demonstrate it once or twice for all to see. Keen observers will soon catch on.*

5. Hit the hoop on the *inside*. This flattens the loop somewhat, pulling it down and away from the foil wad. Inertia at first prevents this wad from moving in any direction, then gravity pulls it straight down into the bottle.

Hitting the hoop on the outside will not work. This pushes the foil wad higher, overcoming its inertia and pushing it away from the bottle.

INSIDE HIT   OUTSIDE HIT

Materials

☐ A 4x6 index card.
☐ Scissors.
☐ Masking tape or clear tape.
☐ A bottle or Erlynmeyer flask.
☐ Aluminum foil.

(TO) examine the distribution of balanced force in a line. To find 4 rubber bands of nearly equal length for later vector analysis.

TENSION O Motion ( )

1. Stretch 12 rubber bands of uniform size to their limit, then string them between 13 unbent paper clips. Stretch this chain across your floor between 2 rubber-banded textbooks so each band stretches to about 3/4 capacity.

a. Are all forces in this system balanced? How do you know?
b. Diagram and label the *horizontal* forces that act on 1 of the paper clips.
c. Diagram and label *all* forces acting on 1 of the books.
d. Does each rubber band have the same tension as it neighbor? Explain.

2. Without moving the books, measure the length of all 12 stretched bands (from left to right), to the nearest .1 cm. Make a data table.

band number	length (cm)
1	
2	
3	
⋮	

3. Select 4 rubber bands with nearly the same length (within 1 cm). Fix each to its paper clip with masking tape, write its length on the tape, and save.

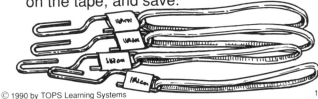

13

Introduction

Roll a sheet of paper into a wide cylinder. Tape the ends so they just overlap, then rest an index card on top. Diagram the force of gravity pushing down, balanced by the cylinder pushing up. Place books on top of the index card, one at a time. After each addition, diagram how the forces increase but are still balanced. Keep increasing the load (and class anticipation) until your cylinder suddenly collapses. Diagram how the forces become unbalanced in free fall, then balanced once more on the table's surface.

NEWTON'S FIRST LAW (version 2)

Mass has *inertia*. This means...
• a body at rest tends to stay at rest *unless acted upon by unbalanced force.*
• a body in motion tends to stay in motion in a straight line *unless acted upon by unbalanced force.*

Answers / Notes

1. *When a new rubber band is first stretched it doesn't quite relax back to its original length. By stretching all 12 rubber bands to their full limit now, you insure that they have uniform stretching characteristics later on.*

1a. All forces are balanced. The chain is not moving.
1b.
1c.

1d. Yes. Tension is evenly distributed throughout the entire line. Each rubber band pulls, and is pulled by, the same amount of force. All forces are balanced.

2-3. *Students should present a table with 12 entries, then tag 4 rubber bands that have nearly equal length.*

Materials

☐ Thin rubber bands of uniform size. Try to match, as closely as possible, this size 16 rubber band drawn to actual size. It works best here, as well as in activities 27 and 28.
☐ Paper clips with uniform size.
☐ Two heavy books and masking tape.
☐ A meter stick or metric ruler.

No. 16 RUBBER BAND
(actual size)

(TO) calibrate the stretch in a rubber band against a laboratory spring scale. To observe that this stretch is roughly linear.

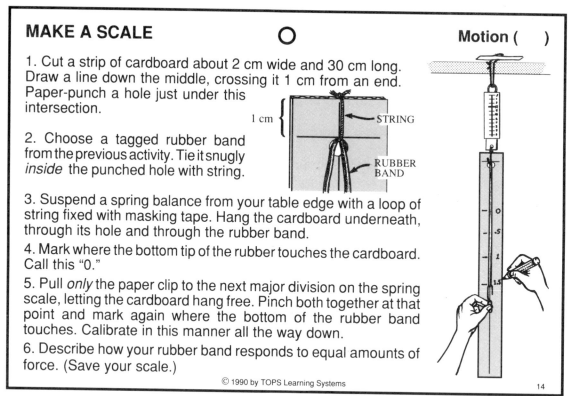

MAKE A SCALE   ○   Motion (   )

1. Cut a strip of cardboard about 2 cm wide and 30 cm long. Draw a line down the middle, crossing it 1 cm from an end. Paper-punch a hole just under this intersection.

1 cm {   STRING

2. Choose a tagged rubber band from the previous activity. Tie it snugly *inside* the punched hole with string.

RUBBER BAND

3. Suspend a spring balance from your table edge with a loop of string fixed with masking tape. Hang the cardboard underneath, through its hole and through the rubber band.

4. Mark where the bottom tip of the rubber touches the cardboard. Call this "0."

5. Pull *only* the paper clip to the next major division on the spring scale, letting the cardboard hang free. Pinch both together at that point and mark again where the bottom of the rubber band touches. Calibrate in this manner all the way down.

6. Describe how your rubber band responds to equal amounts of force. (Save your scale.)

14

Answers / Notes

4. *The zero point on the cardboard scale should correspond to zero on the spring scale. Some spring scales may read slightly higher than this since the cardboard, rubber band and paper clip do weigh a small but significant amount. If your spring scales have adjustable zero points, use them to center in on true zero for greater accuracy.*

5. *All force must be applied only to the rubber band, while the cardboard scale "floats" behind. Try this important step yourself, before students try it. If necessary, demonstrate this process to your whole class. The rubber band scales calibrated here will be used in the next 2 activities.*

6. Initally, the rubber band stretches by increasing amounts when equal increments of force are applied. After about 1.5 N, the distance intervals become regular. That is, the rubber band stretches in direct proportion to the applied force.

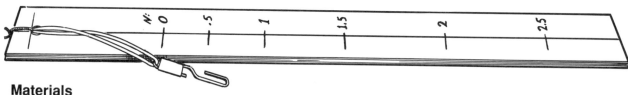

Materials

☐ A cardboard strip. You may wish to precut these 2 cm by 30 cm strips on a paper cutter. For maximum strength and flatness, always cut the long dimension so it runs parallel to the corrugations.

☐ Provide heavy-duty scissors and metric rulers if you wish students to cut their own strips from larger pieces of cardboard. Again, have them cut the long dimension parallel to the corrugations.

☐ A paper punch.

☐ A tagged rubber band and paper clip from the previous activity.

☐ String and scissors.

☐ A spring scale with a 2.5 N (250 g) capacity. Don't substitute scales under 2 N (200 g) or over 3 N (300 g). As an alternative to spring scales, students can also calibrate their rubber band scales directly against mass by adding 50 g (.5 N) increments to attached paper cups.

☐ Masking tape.

(TO) compare static friction with moving friction. To observe how static friction balances an applied force in an equal and opposite direction.

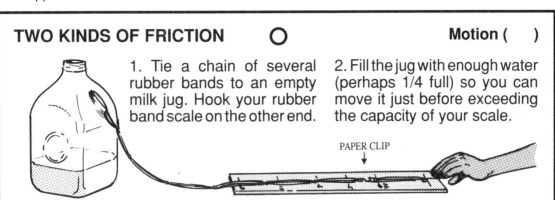

TWO KINDS OF FRICTION   O   Motion (   )

1. Tie a chain of several rubber bands to an empty milk jug. Hook your rubber band scale on the other end.

2. Fill the jug with enough water (perhaps 1/4 full) so you can move it just before exceeding the capacity of your scale.

PAPER CLIP

3. Compare the force required to *start* the jug moving (static friction) with the force required to *keep* it sliding (moving friction).
    a. Are these 2 kinds of friction overcome by the same amount of force? Support your answer with numbers.
    b. Propose a hypothesis to explain your observations.

4. As you pull the scale with more and more force, rubber bands on the jug pull in an equal and opposite direction against you!
    a. What evidence can you observe to support this idea?
    b. Why does the jug eventually move? Write about balanced and unbalanced forces as you answer.

© 1990 by TOPS Learning Systems   15

Introduction

Push against a wall. Ask your class how they might demonstrate that the wall is pushing back with a balanced force: (a) Someone might insert their own hand between your hand and the wall, and feel equal pressure on both sides. (b) Discuss what would happen if the wall didn't push back with an equal force! (c) Push against another student who is pushing back with equal force, then suddenly withdraw your hand.

Answers / Notes

3a. Answers will vary widely. Nevertheless, all students should report that static friction is larger than moving friction. More force is required to *start* the jug moving than to *keep* the jug moving once it is underway.

3b. There are two explanations: (1) A certain amount of unbalanced force is required to overcome the inertia of the milk jug and set it in motion. (2) The surfaces are not completely smooth. Microscopic ridges and valleys in the table settle into corresponding valleys and ridges on the bottom of the milk jug. Static surfaces can actually lock together somewhat, producing greater friction than when the surfaces slide continually past each other.

4a. The rubber bands on the jug handle stretch more tightly in direct response to your pull on the scale.

4b. As long as static friction resists motion in an equal and opposite direction to the pull of the scale, the jug won't move. Its inertia keeps it at rest because the forces are balanced. Eventually, the pull of the scale exceeds static friction between the jug and table. The jug moves in response to this unbalanced force in the direction of the scale.

Materials

☐ Rubber bands.
☐ An empty gallon milk jug with a handle.
☐ A source of water.
☐ The rubber band scale constructed in the previous activity. A commercial spring balance may be substituted. Be aware that it may not be designed to measure accurately in a horizontal position. An increase in internal friction might causes it to register too little force.

(TO) study angles and magnitudes of opposing force vectors.

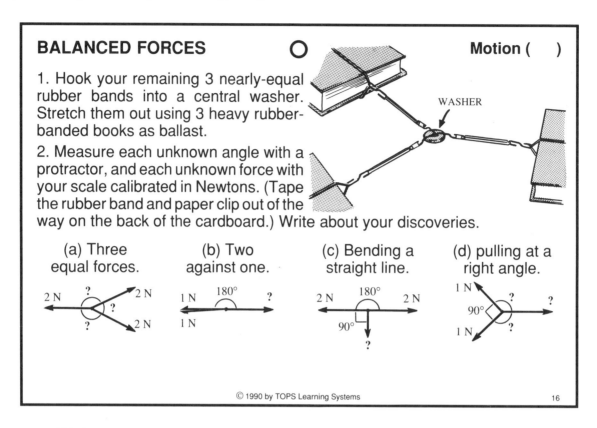

BALANCED FORCES ◯   Motion ( )

1. Hook your remaining 3 nearly-equal rubber bands into a central washer. Stretch them out using 3 heavy rubber-banded books as ballast.

WASHER

2. Measure each unknown angle with a protractor, and each unknown force with your scale calibrated in Newtons. (Tape the rubber band and paper clip out of the way on the back of the cardboard.) Write about your discoveries.

(a) Three equal forces.

(b) Two against one.

(c) Bending a straight line.

(d) pulling at a right angle.

16

Answers / Notes

2. *Recall that the 3 rubber bands linked to the washer were specifically chosen to stretch by the same distance as the calibrated rubber band used in the Newton force scale. This scale, therefore, makes a perfect Newton "ruler," as long as your class understands how to use it properly: the top of the hole should be aligned with one end of the stretched rubber band, and the correct force read where the other end of the rubber band meets the scale.*

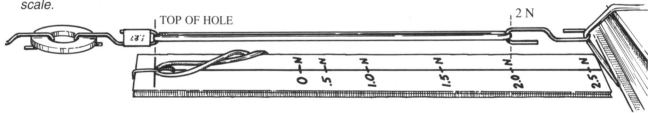

(a) 119 °, 121 °, 121 °
Given 3 vectors pulling with the same force and lying in the same plane, the angles between them equal 120 ° and add to 360 °.

(b) 2.1 N
One vector pulling in direct opposition to 2 others must have twice the magnitude to balance.

(c) 0 N (no force)
Two forces pulling in equal and opposite directions create a straight angle of 180 ° between them. *Any* force applied at a right angle to these two forces will significantly bend this straight line.

(d) 1.4 N, 135 °, 135°
This force is exactly √2 times longer than either of the other 2 forces. This follows because the two equal forces form 2 sides of a square that resolve into its diagonal.

Materials

☐ Tagged rubber bands and paper clips from activity 13.
☐ A washer.
☐ Three heavy books, rubber bands.
☐ A protractor.
☐ The rubber-band scale from activity 14.

(TO) build a simple acceleration indicator and observe how it responds to changes in speed and direction.

BUILD AN ACCELEROMETER ◯ Motion (   )

1. Fix a paper clip and thread with masking tape under the lid of a baby food jar, so it hangs in the center when closed, not quite touching the bottom.

TAPE (inside lid)

THE CLIP CAN SWING FREELY.

2. Fill the jar with water and tightly close the lid.

3. Push your accelerometer so it slides untouched across a table and comes to rest. (Don't let it fall off the edge!) Explain in words and pictures how the paper clip shows acceleration (speeding up); deceleration (slowing down).

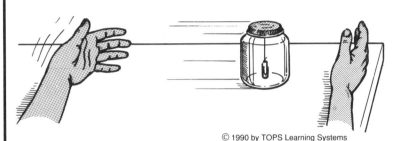

4. Walk about the room holding your accelerometer. What can you discover?

© 1990 by TOPS Learning Systems   17

Answers / Notes

3. As the jar accelerates forward from the push of your hand, the paper clip pendulum inside leans backward, into the applied force. Being at rest, the paper clip tends to stay at rest.

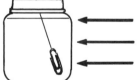

EARLY ACCELERATION

As friction decelerates that jar to a sliding stop, the paper clip pendulum inside leans forward, into the frictional force. Being in motion, the paper clip tends to stay in motion.

LATER DECELERATION

4. *Observations in this step will be examined again in the next activity. Student responses might include some, perhaps not all, of the following:*

    a. When increasing speed or starting forward from a standstill, the paper clip moves back. (Acceleration.)

    b. When decreasing speed or stopping, the paper clip moves forward. (Deceleration or negative acceleration.)

    c. When moving at constant speed in a straight line, the paper clip hangs straight down. (No acceleration.)

    d. When moving at a constant speed through a curved path or turning in a circle, the paper clip leans out toward the arc. (Acceleration.)

Materials

☐ A paper clip.
☐ Thread.
☐ A baby food jar with lid.
☐ Scissors.
☐ Masking tape.
☐ A source of water.

(TO) understand acceleration as a change in speed or direction.

ACCELERATION IS CHANGE! ○   Motion ( )

ACCELEROMETER

In each case below…
  a. Identify the acceleration acting on your accelerometer (if any), naming the balanced or unbalanced forces involved.
  b. Draw a diagram showing force vectors.

1. Push the jar to *start it moving.*

2. Allow a moving jar to *come to a free sliding rest.*

3. Move the jar across your table *in a straight line, at constant speed.* (Maintain continuous contact with your hand.)

4. Observe the jar *at rest.*

5. Move the jar on your table *in a circle at constant speed.* (Diagram this with a top view.)

© 1990 by TOPS Learning Systems   18

Introduction

Summarize Newton's second law of motion on your blackboard.

NEWTON'S SECOND LAW (version 1)
An object accelerates if acted upon by unbalanced force. This changes its speed, direction or both.

Wad up a piece of paper and set it on the table. Ask why it isn't moving. (Objects at rest tend to stay at rest unless acted upon by an unbalanced force.) Then throw it to a student. Ask why it moved. (It was accelerated by an unbalanced force — throwing it.) Ask why it didn't move in a straight line. (It was accelerated down by an unbalanced force — gravity.) Ask why it stopped. (It was decelerated by an unbalanced force — catching it.)

Answers / Notes

1. The jar accelerates. Its speed increases in response to an unbalanced push.

2. The jar decelerates. Its speed decreases in response to unbalanced friction.

3. The jar does not accelerate. It changes neither speed nor direction. The force of friction is balanced in an equal and opposite direction by the hand.

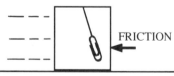

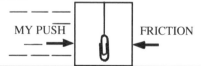

4. The jar does not accelerate. It changes neither speed nor direction. No unbalanced forces act upon it. Gravity pushes it down, but this force is balanced in an equal and opposite direction by the table pushing up.

ALL FORCES BALANCE

5. The jar accelerates. Its direction (not speed) changes in response to the unbalanced restraining force of the hand, which keeps it from moving in a straight line.

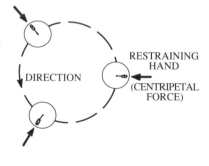

Materials

☐ The accelerometer constructed in the previous activity.

(TO) describe speed and acceleration in numbers and units. To correlate motion with time vs. distance graph lines.

SECONDS PER SECOND ○ Motion ( )

1. Each square below represents 1 meter that you travel on a bicycle. Fully describe your motion (in numbers plus units) from left (t = 0 sec) to right. The first "voyage" is done as an example.

a. →
-1 0 1 2 3 4 5 6 7 8 9 10 SECONDS

(You are moving at a constant speed of 1 m/sec. At t = 7 sec, you begin to accelerate 2 m/sec each new second, or 2 m/sec².)

b. st
0 1 2 3 4 →

c. st
0 1 2 3 4 5 6 7 8 9 end

d. →
-1 0 1 2 3 4 5 6,7,8,9 / 10 11 12 13

2. Graph and label each "voyage" on the same pair of coordinates, beginning at (0,0). Check your descriptions against each graph line.

DIST (m) / TIME (sec)

19

Answers / Notes

1a. (As above.)

1b. Starting from rest, you accelerate at 3m/sec² throughout.

1c. Starting from rest you accelerate 1 m/sec² until t = 5 sec. Then you decelerate 1 m/sec² until coming to rest once more.

1d. You are moving at a constant speed of 2 m/sec, then decelerate to a dead stop between the 5th and 6th second. You remain at rest for 3 seconds, then accelerate once agains at 1 m/sec².

2.

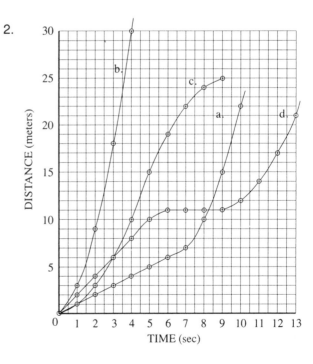

Materials

☐ Graph paper.

(TO) track a marble as it is accelerated by gravity on an inclined plane. To understand why the marble follows a parabolic curve.

TRACKING A CURVE ◯ Motion (   )

1. Lean a manila folder on a book. Bridge the gap at the bottom with scratch paper.

2. Slide your index card ramp up to the corner of this incline so a marble rolls up and back down, traveling across most of the paper. (Set the ramp on a lump of clay if necessary.)

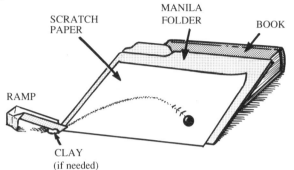

3. Dip a marble in soapy colored water and let it trace a path across your paper.

4. Identify where your marble accelerates and decelerates.

5. The curved path followed by the marble is called a *parabola*.
   a. If the marble has inertia, why doesn't it travel in a straight line with uniform motion?
   b. If you throw a stone will it trace out a similar parabola? Explain.

20

Answers / Notes

4.

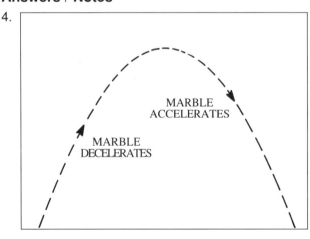

5a. The marble follows a curved parabolic path because it is continuously acted upon by the unbalanced force of gravity.

5b. Yes. Gravity decelerates the rock as it rises, turns it around, and accelerates it downward. This compromise between inertia and gravity results in a parabola.

Materials

☐ A manila file folder. A sheet of tagboard or thin cardboard will also serve.
☐ A book.
☐ Scratch paper.
☐ Index card ramps from activities 8 and 9.
☐ A marble.
☐ A lump of clay.
☐ Soapy water tinted with food coloring in a labeled jar.
☐ Jar lids or crucibles to hold the tracking solution.

(TO) develop a model that illustrates the interplay of inertia and gravity in orbiting satellites.

EARTH-MOON MODEL ○ Motion ( )

1. Tie a paper clip to both ends of a 30 cm thread that has been pulled through a 10 cm straw. Add 5 additional paper clips to one end.

2. Spin the single paper clip just fast enough to support the other 6 clips about 1 cm from the bottom of the straw.

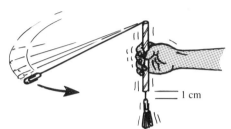

3. Let this system model the earth and moon.

    a. What represents the moon? the earth? gravity?
    b. Why doesn't the moon come crashing to earth?
    c. If the moon's orbit slowed, would it be pulled straight to earth? Explain.
    d. Suppose Earth's moon had twice the mass. Would it circle the earth at the same rate? Back up your answer with experimental evidence.

21

Answers / Notes

3a. The single revolving paper clip represents the moon; the 6 paper clips and straw stand for the earth *(with 6 times as much mass);* the thread models the gravitational pull between the earth and the moon.

3b. The moon's orbit is a compromise between its inertial tendency to move in a straight line and its acceleration toward earth by the unbalanced force of gravitational attraction.

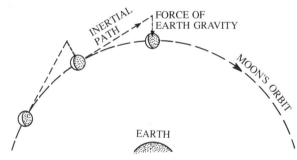

3c. No. As demonstrated by the paper clip model, when the moon's orbit slows, gravity pulls it into increasingly tighter and faster orbits until it spirals into the earth.

3d. No, the moon would circle the earth more slowly. *(Students should support their answer by timing and comparing the frequency of revolutions for 1 and 2 paper-clip moons. A single paper clip, for example, might revolve 37 times in 10 seconds, compared to 2 paper clips revolving 25 times in 10 seconds.)*

Materials

☐ Paper clips.
☐ Thread.
☐ A metric ruler.
☐ A plastic straw. Glass tubing with smooth fire-polished ends works even better.

(TO) understand Newton's second law of motion — that acceleration increases in direct proportion to force and inverse proportion to mass.

FORCE OVER MASS... ○   Motion (   )

1. Push your accelerometer using different amounts of unbalanced force. Summarize your results in a table.

unbalanced force	acceleration
small push	
medium push	
large push	

2. Newton recognized that acceleration is *directly proportional* to force. How do your results support his observation?

3. Fill 2 more baby food jars with water and close the lids. Practice applying equal force, first to 1 jar, then 2, then 3 in a row. Do this by pushing as hard as you can, using only the *back* of your little finger, and keeping your wrist on the table.

WRIST DOWN

4. Summarize your results in a table. How do your results support Newton's observation that acceleration is *inversely proportional* to mass?

mass	acceleration
small (1 jar)	
medium (2 jars)	
large (3 jars)	

5. Newton's second law is summarized by this equation: What does it mean?

$$a = \frac{F}{m}$$

© 1990 by TOPS Learning Systems   22

Introduction

Summarize Newton's second law of inertia on your blackboard.

> **NEWTON'S SECOND LAW (revised)**
> Objects accelerate...
> • in direct proportion to the applied force,
> • in inverse proportion to their mass.

This law can be translated mathematically as acceleration = constant x force/mass. Substituting Newtons for force (F), kilograms for mass (m), and meters per second squared for acceleration (a), the constant in this equation reduces to 1, and the equation simplifies to: $a = F/m$. This means that...

Doubling **F** doubles **a**:

1 N → [1 kg]   2 N ⟶ [1 kg]   4 N ⟶ [1 kg]
 (a = 1/1)   (a = 2/1)   (a = 4/1)

Doubling **m** halves **a**:

1 N → [1 kg]   1 N → [2 kg]   1 N → [4 kg]
 (a = 1/1)   (a = 1/2)   (a = 1/4)

Answers / Notes

1-2.

unbalanced force	acceleration
small push	small
medium push	medium
large push	large

As force increases, acceleration also increases.

4.

mass	acceleration
small (1 jar)	large
medium (2 jars)	medium
large (3 jars)	small

As mass increases, acceleration decreases.

3. *It is not possible to generate a whole lot of force with the back of your little finger. The idea here is that the* maximum *application of force applied in each case will result in a more or less* constant *application of force.*

5. This equation states that acceleration increases in direct proportion to force and decreases in inverse proportion to mass.

Materials

☐ The accelerometer constructed in activity 17.
☐ Two baby-food jars with lids, plus water.

(TO) calculate the acceleration of gravity on 3 different masses. To observe that the result is always the same, roughly 10 m/sec² on Earth.

FREE FALL   ◯   Motion (   )

1. Make 3 penny weights like these. Hold them together with masking tape.

1 PENNY

2 PENNIES

4 PENNIES

2. Gravity pushes down on 1 penny with a force of .025 Newtons. (This is its *weight*.) Calculate the weight of 2 pennies; of 4 pennies.

3. One penny has a *mass* of .0025 kg. Calculate the mass of 2 pennies; of 4 pennies.

4. In free fall, an object is pulled down by the force of its own weight. Calculate how fast each mass accelerates in free fall.

$$a = \frac{F}{m}$$

5. People commonly think that heavy objects fall faster than light objects.
   a. What do your results in step (4) say about this?
   b. Drop all 3 penny weights together and write what you observe.

23

Introduction

One Newton is defined as the force required to accelerate 1 kg of mass exactly 1 meter per second faster, each new second. It follows from this definition that 1 Newton = 1 kg·m/sec/sec. From a units perspective, Newton's second law makes good sense: force divided by mass yields units of acceleration.

$$a = \frac{F}{m} = \frac{N}{kg} = \frac{kg \cdot m/sec^2}{kg} = m/sec^2$$

Answers / Notes

2.   weight of 2 pennies = 2 x .025 N = .05 N
     weight of 4 pennies = 4 x .025 N = .10 N

3.   mass of 2 pennies = 2 x .0025 kg = .005 kg
     mass of 4 pennies = 4 x .0025 kg = .010 kg

4. $a = \frac{F}{m} = \frac{.025\,N}{.0025\,kg} = 10\ m/sec^2$ $a = \frac{F}{m} = \frac{.05\,N}{.005\,kg} = 10\ m/sec^2$ $a = \frac{F}{m} = \frac{.1\,N}{.01\,kg} = 10\ m/sec^2$

5a. Because the light, medium, and heavy penny weights all accelerate at the same 10 m/sec², they all fall together. *(This is true of any object. As its weight increases, so does its inertial mass. Acceleration, the ratio of weight to mass, thus stays constant.)*

5b. The 3 penny-weights fall together and hit the floor at the same time.

Materials

☐ Pennies.
☐ Masking tape.

(TO) understand why lighter objects fall through air more slowly than heavier objects.

AIR RESISTANCE ⭘ Motion ( )

1. Divide a sheet of scratch paper exactly in half. Crumple one, leave the other flat.

DROP TOGETHER.

a. Compare their rate of free-fall.

b. What is the real cause of objects not falling together? Draw a labeled diagram to show the forces acting on each paper.

2. Crumple the flat sheet around 10 pennies until it is the same size as the other.

DROP TOGETHER.

10 PENNIES INSIDE

a. Now that they have the same air resistance, do they free-fall together?

b. What role does mass play in overcoming air resistance? Explain.

3. Put a penny and a wisp of cotton (or a feather) in a paper cup.
   a. Predict what will happen when you drop the cup. Refer to Newton's second law.   b. Test your prediction.

24

Answers / Notes

1a. The crumpled paper falls much faster than the flat sheet.

1b. Objects with more air resistance fall slower than objects with less air resistance.

WEIGHT

AIR RESISTANCE

2a. No. The paper crumpled around 10 pennies falls slightly faster than the lighter crumpled paper.

2b. As an object's mass increases, so does its ability to overcome air resistance. This is because more massive objects have greater inertia. Once accelerated into motion they are slowed less by air resistance.

3a. Newton's second law predicts that the ratio of weight to mass (acceleration) is the same for all falling objects. The penny and the wisp of cotton should fall together inside the cup because neither is slowed by air resistance. Only the cup will be slowed by air resistance. It will slow the acceleration of the penny and cotton just a little, causing them to press lightly against the bottom of the cup as it falls.

3b. Both the cotton and the penny remained in the bottom of the cup as it fell to the floor.

Materials

☐ Scratch paper.
☐ Pennies.
☐ Cotton or a feather.
☐ A paper cup.

(TO) interpret an action-reaction event in terms of Newton's third law of motion.

CLOTHESPIN LAUNCHER ○   Motion (   )

1. Fix a clothespin to a baby food jar with two rubber bands so the jaws point up.

2. Wedge the clothespin wide open with another half clothespin tied to a trip thread.

3. Put this next to a baby food jar lid in an open area on the floor. Pull the thread with a *quick* flick of your wrist. Describe what happens.

4. Interpret your observations in terms of Newton's third law.

HALF CLOTHESPIN (turn sideways)

JAWS UP

TRIP THREAD

LID

5. According to Newton's second law, the acceleration of any object is inversely proportional to its mass. Demonstrate this by pressing a lump of clay into the lid. What can you conclude?

25

Introduction

Summarize Newton's third law of motion on your blackboard.

> **NEWTON'S THIRD LAW**
> When one body exerts a force upon a second body, the second body exerts an *equal* and *opposite* force upon the first.

Identify action/reaction experiences that are familiar to your students. Discuss the equal and opposite forces involved:

(a) Pushing a car: Shove a car forward and it pushes your body back with equal force.
(b) Rowing a boat: Push the water ahead and it pushes the boat back with equal force.
(c) Shooting a gun: As the exploding gun powder forces the bullet forward, it kicks the gun backward with equal force.

Answers / Notes

3. The clothespin "wings" snap open, accelerating the lid and jar in opposite directions. The lid travels much farther than the jar.

4. When the clothespin snaps , it exerts a force on the lid. As Newton's third law predicts, the lid exerts an equal and opposite force on the clothespin and attached jar. Both objects, therefore, accelerate away from each other in opposite directions.

5. Adding more and more clay to the lid steadily reduces the distance it slides after being snapped by the clothespin. Since the clothespin snaps with essentially the same force whether the lid is full or empty, it must be the mass of the lid that is affecting how far it moves. In accordance with Newton's second law ($a = F/m$), this mass is inversely proportional to acceleration: smaller masses experience greater acceleration and therefore travel longer distances; larger masses experience less acceleration and therefore travel shorter distances.

Materials

☐ Clothespins.
☐ A baby food jar with lid.
☐ Rubber bands.
☐ Thread.
☐ Modeling clay.

(TO) build a rotating jet balloon. To understand its motion in terms of Newton's third law.

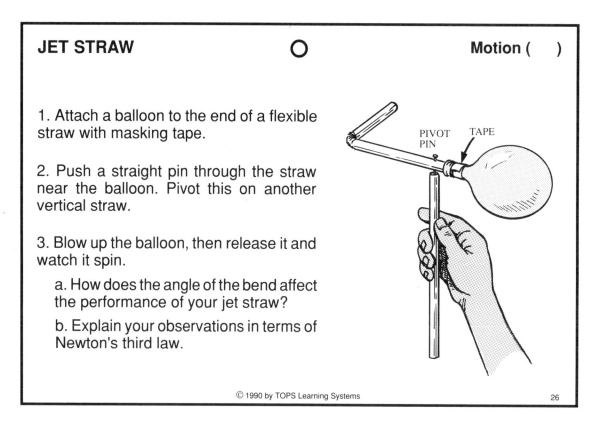

JET STRAW   O   Motion ( )

1. Attach a balloon to the end of a flexible straw with masking tape.

2. Push a straight pin through the straw near the balloon. Pivot this on another vertical straw.

3. Blow up the balloon, then release it and watch it spin.

    a. How does the angle of the bend affect the performance of your jet straw?

    b. Explain your observations in terms of Newton's third law.

26

Answers / Notes

3a. The straw turns fastest when bent at a right angle. Straightening this bend causes the straw to turn slower. When the straw is perfectly straight, it won't spin at all.

3b. The bend in the straw forces the moving air to turn a corner. This accelerating air stream, in accordance with Newton's third law, pushes back against the constraining bend with equal and opposite force. Because this force is directed at a right angle to the pin, the straw spins in a circle around it. Straightening the straw redirects the air stream to push more directly against the pin rather than around it. The straw no longer moves as rapidly because the pin is constrained.

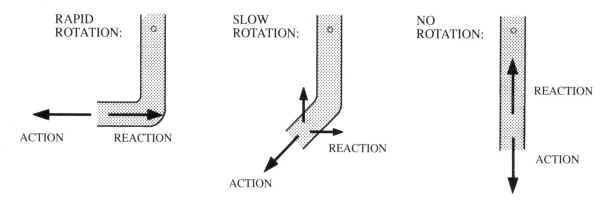

Materials

☐ A balloon.
☐ Flexible plastic drinking straws.
☐ Masking tape.
☐ A straight pin.

(TO) construct a rubber band catapult for use in a study of force and mass.

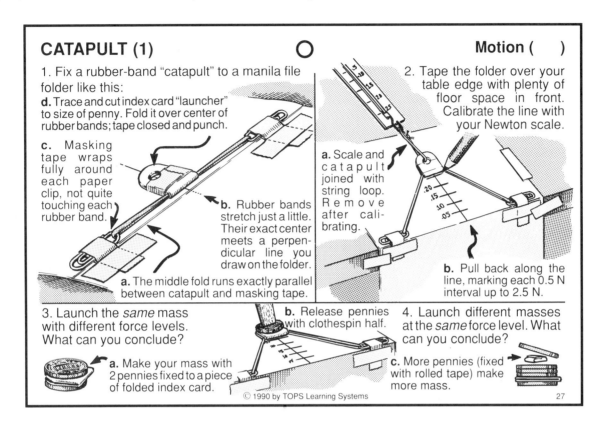

CATAPULT (1) ○

1. Fix a rubber-band "catapult" to a manila file folder like this:

d. Trace and cut index card "launcher" to size of penny. Fold it over center of rubber bands; tape closed and punch.

c. Masking tape wraps fully around each paper clip, not quite touching each rubber band.

b. Rubber bands stretch just a little. Their exact center meets a perpendicular line you draw on the folder.

a. The middle fold runs exactly parallel between catapult and masking tape.

Motion ( )

2. Tape the folder over your table edge with plenty of floor space in front. Calibrate the line with your Newton scale.

a. Scale and catapult joined with string loop. Remove after calibrating.

b. Pull back along the line, marking each 0.5 N interval up to 2.5 N.

3. Launch the *same* mass with different force levels. What can you conclude?

a. Make your mass with 2 pennies fixed to a piece of folded index card.

b. Release pennies with clothespin half.

4. Launch different masses at the *same* force level. What can you conclude?

c. More pennies (fixed with rolled tape) make more mass.

© 1990 by TOPS Learning Systems   27

Answers / Notes

2. *By fixing the catapult to the file folder, and then taping this folder to the table, you create a mobile launcher. It can easily be removed on a temporary basis, or placed somewhere else as the need arises.*

1d, 3a. *Both steps require elaboration. On an index card, trace twice around a penny, leaving about a penny-thickness of space between. Then cut and fold. In step 3a only, fix a penny to each outside half using masking tape rolled sticky side out. Use this as your standard launching mass.*

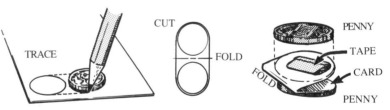

TRACE   CUT   FOLD   PENNY   TAPE   CARD   FOLD   PENNY

3. As the 2 pennies are launched with more force, they experience greater acceleration, and thus travel longer distances in accordance with Newton's second law. *(It should be noted that pennies launched at a particular force level, 2.0 N for example, are only initially accelerated by this force. It drops from 2.0 N down to 0 N over the 10 cm-or-so interval, beginning at the point of release and ending as the penny leaves the table edge.)*

4. As more mass (more pennies) is launched at the same force level, it experiences less acceleration, and thus covers shorter distances. Again, this is in accordance with Newton's second law.

Materials

☐ Rubber bands. Use the same size-16 rubber band as illustrated in activity 13. Two of these chained together will span from 12 to 14 cm when stretched just a little. Where they join defines the center of the catapult. This length provides enough force to launch the pennies over significant distances that will graph well in activity 28. But the rubber bands won't be long enough to turn the flying pennies into dangerous projectiles. If your particular rubber bands are too long or too short, substitute fewer or more to span the desired 12 cm to 14 cm distance.

☐ A manila file folder.　☐ An index card.
☐ Paper clips.　☐ The Newton scale constructed in activity 14.
☐ Masking tape.　☐ String and scissors.
☐ A paper punch.　☐ Pennies.

(TO) graph how acceleration is directly proportional to force and inversely proportional to mass.

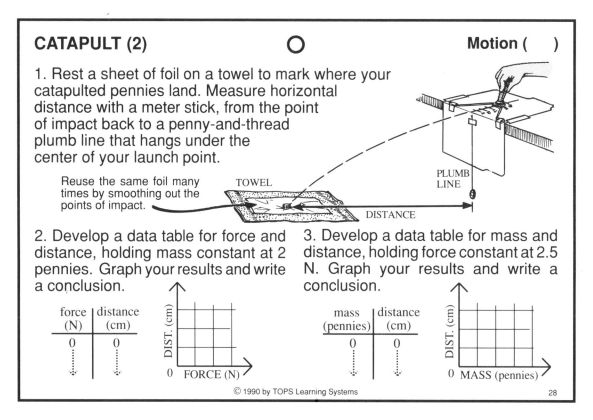

CATAPULT (2)   O   Motion (   )

1. Rest a sheet of foil on a towel to mark where your catapulted pennies land. Measure horizontal distance with a meter stick, from the point of impact back to a penny-and-thread plumb line that hangs under the center of your launch point.

Reuse the same foil many times by smoothing out the points of impact.

TOWEL   PLUMB LINE   DISTANCE

2. Develop a data table for force and distance, holding mass constant at 2 pennies. Graph your results and write a conclusion.

force (N)	distance (cm)
0	0
⋮	⋮

DIST. (cm)   0 FORCE (N)

3. Develop a data table for mass and distance, holding force constant at 2.5 N. Graph your results and write a conclusion.

mass (pennies)	distance (cm)
0	0
⋮	⋮

DIST. (cm)   0 MASS (pennies)

© 1990 by TOPS Learning Systems   28

Answers / Notes

2-3. *These graph lines resulted from measuring each distance over 3 trials and striking an average.*

The straight graph line suggests that the initial force that accelerates both pennies increases in direct proportion to the horizontal distance they travel (a measure of their original acceleration within the catapult). This confirms Newton's observation that force and acceleration increase in direct proportion.

The curved graph line suggests that as the number of pennies increases, the horizontal distance they travel (a measure of their original acceleration) decreases in an inverse proportion. This confirms Newton's observation that acceleration is inversely proportional to mass.

Extension

All pennies fall through the same vertical distance, from table edge to floor, no matter how far they travel horizontally. Calculate the time this takes ($d = 1/2gt^2$). Knowing this, calculate the speed required for 2 pennies to travel 2 meters ($v = s/t$). Knowing this, calculate their kinetic energy ($KE = 1/2mv^2$). Knowing this, calculate the work put into the catapult ($W_{in} = W_{out}$). Now plot the force you pull vs. the distance the launcher moves back. (Begin measuring at a point where the rubber band is *just* able to push the penny over the table edge.) Find how much area under your curve equals W_{in}. (Count graph-paper squares.) Pull your catapult back a corresponding distance to see if the 2 pennies travel the predicted 2 meters.

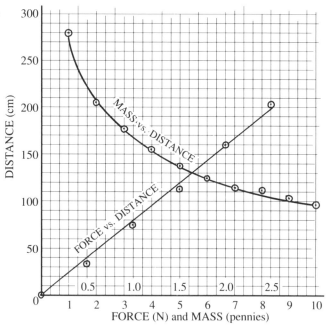

MASS vs. DISTANCE
FORCE vs. DISTANCE

DISTANCE (cm)

300
250
200
150
100
50
0

0.5   1.0   1.5   2.0   2.5

1 2 3 4 5 6 7 8 9 10
FORCE (N) and MASS (pennies)

Materials

☐ The catapult constructed in the previous activity.
☐ Aluminum foil.
☐ A towel.
☐ Masking tape.
☐ Thread and pennies.
☐ A meter stick.

(TO) listen to paper clips on a string as they fall to the floor. To relate the frequency of these taps to the acceleration of gravity.

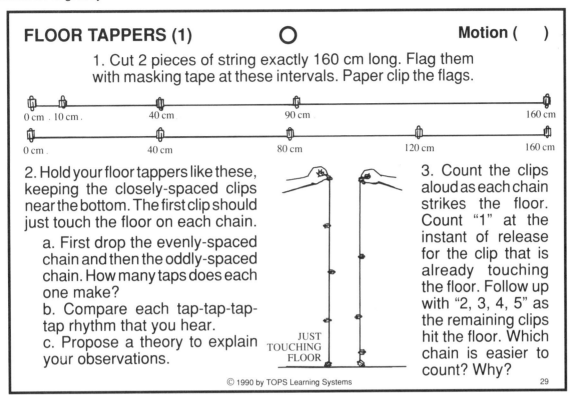

FLOOR TAPPERS (1)   O   **Motion (   )**

1. Cut 2 pieces of string exactly 160 cm long. Flag them with masking tape at these intervals. Paper clip the flags.

0 cm . 10 cm .   40 cm   90 cm .   160 cm

0 cm .   40 cm   80 cm   120 cm   160 cm

2. Hold your floor tappers like these, keeping the closely-spaced clips near the bottom. The first clip should just touch the floor on each chain.

    a. First drop the evenly-spaced chain and then the oddly-spaced chain. How many taps does each one make?
    b. Compare each tap-tap-tap-tap rhythm that you hear.
    c. Propose a theory to explain your observations.

JUST TOUCHING FLOOR

3. Count the clips aloud as each chain strikes the floor. Count "1" at the instant of release for the clip that is already touching the floor. Follow up with "2, 3, 4, 5" as the remaining clips hit the floor. Which chain is easier to count? Why?

29

Answers / Notes

2a. Each floor tapper produces 4 distinct taps. The first paper clip on each chain doesn't sound because it is already resting on the floor.

2b. The evenly-spaced chain produces a rhythm of taps that sound faster and faster. The oddly-spaced chain produces a rhythm of taps that sound evenly.

          evenly-spaced tapper: tap———tap——tap—tap.
          oddly-spaced tapper: tap——tap——tap——tap.

2c. Gravity accelerates each chain of paper clips in free fall, causing it to fall faster and faster. The paper clips on the evenly-spaced chain thus strike the floor at more frequent intervals as the chain speeds up. The paper clips on the oddly-spaced chain also speed up, but each succeeding clip also travels a greater distance than the one before. These increasing speeds and distances match to produce equal time intervals between the taps (a steady beat).

3. The oddly-spaced chain is easier to count. Due to the increasing space because paper clips, the accelerating chain still strikes the floor in a regular rhythm. Gravity accelerates the evenly-spaced chain, by contrast, until the last paper clips land almost too rapidly to count.

Materials

☐ A meter stick.
☐ String.
☐ Scissors.
☐ Masking tape.
☐ Paper clips.
☐ A floor surface that is hard enough to produce a distinct tapping sound when you drop a paper clip. If your floor is covered by a rug, or otherwise too soft, a flat piece of corrugated cardboard provides the perfect sounding surface.

(TO) graph the distance through which paper clips free fall as a function of time. To calculate speed and acceleration.

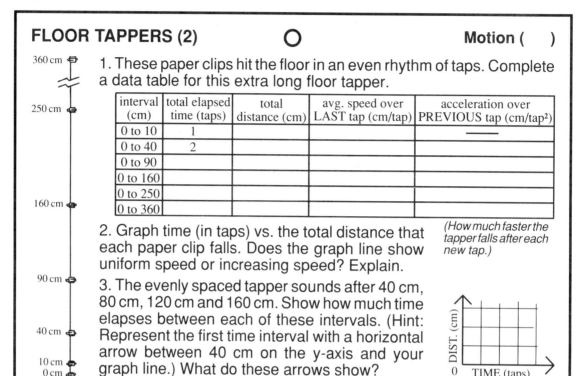

FLOOR TAPPERS (2)　　　○　　　　Motion (　)

1. These paper clips hit the floor in an even rhythm of taps. Complete a data table for this extra long floor tapper.

interval (cm)	total elapsed time (taps)	total distance (cm)	avg. speed over LAST tap (cm/tap)	acceleration over PREVIOUS tap (cm/tap²)
0 to 10	1			——
0 to 40	2			
0 to 90				
0 to 160				
0 to 250				
0 to 360				

2. Graph time (in taps) vs. the total distance that each paper clip falls. Does the graph line show uniform speed or increasing speed? Explain.

3. The evenly spaced tapper sounds after 40 cm, 80 cm, 120 cm and 160 cm. Show how much time elapses between each of these intervals. (Hint: Represent the first time interval with a horizontal arrow between 40 cm on the y-axis and your graph line.) What do these arrows show?

(How much faster the tapper falls after each new tap.)

© 1990 by TOPS Learning Systems

30

Answers / Notes

1.

interval (cm)	total elapsed time (taps)	total distance (cm)	avg. speed over LAST tap (cm/tap)	acceleration over PREVIOUS tap (cm/tap²)
0 to 10	1	10	10	——
0 to 40	2	40	30	20
0 to 90	3	90	50	20
0 to 160	4	160	70	20
0 to 250	5	250	90	20
0 to 360	6	360	110	20

2-3.

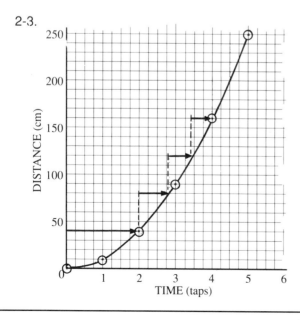

The graph line curves upward, demonstrating that the speed is ever increasing. Over each new regular tap of time, the chain accelerates through increasing distance intervals.

The arrows decrease in length, showing that the evenly-spaced clips beat out smaller and smaller time intervals as the free-falling chain accelerates.

Materials

☐ Graph paper.
☐ Students may need to use both floor tappers from the previous activity as a reference.

(TO) graph how free-falling paper clips accelerate as a function of time. To recognize that the acceleration of gravity is constant.

FLOOR TAPPERS (3)　　○　　　　　Motion (　)

1. Graph the average *speed* of each paper clip over its last interval, just before it hits the floor. (Use the 2nd and 4th columns from the data table in the previous activity.)

> *Note: Shift each point 2 squares (1/2 tap) left, since the average speed happens half-way through each interval.*

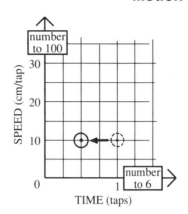

2. Does the graph line show uniform acceleration? Explain.

3. Draw vertical arrows on your graph to show the increase in speed (acceleration) over successive time intervals. Compare your answer with the previous data table.

31

Answers / Notes

1.

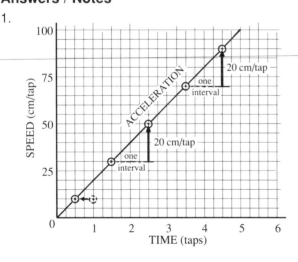

2. Yes. The straight graph line indicates that speed is increasing at a constant rate per unit of time. In other words, acceleration is constant.

3. The arrows show a uniform acceleration of 20 cm/tap^2. That is, the speed increases 20 cm/tap after each new tap. This agrees with column 5 from the data table in the previous activity.

Extension

Use a stopwatch to find the value of a "tap" in seconds. (A paper clip takes .57 seconds to free fall though a distance of 1.6 m or 4 taps. Thus 1 tap = .57 sec / 4 ≈ .14 sec.)

Materials

☐ Data table from the previous activity.
☐ Graph paper.

(TO) compare acceleration down an incline with free-fall.

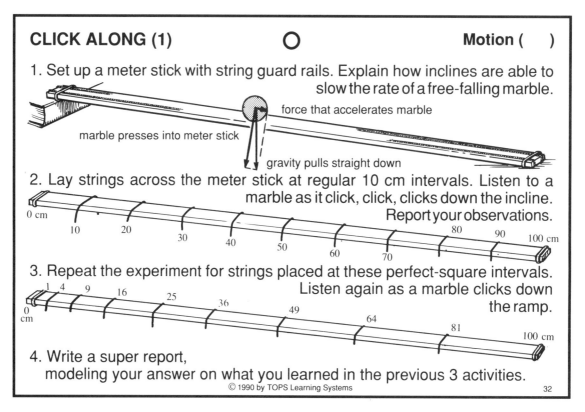

CLICK ALONG (1)　　　　　O　　　　　　**Motion (　)**

1. Set up a meter stick with string guard rails. Explain how inclines are able to slow the rate of a free-falling marble.

force that accelerates marble

marble presses into meter stick

gravity pulls straight down

2. Lay strings across the meter stick at regular 10 cm intervals. Listen to a marble as it click, click, clicks down the incline. Report your observations.

0 cm 10 20 30 40 50 60 70 80 90 100 cm

3. Repeat the experiment for strings placed at these perfect-square intervals. Listen again as a marble clicks down the ramp.

0 cm 1 4 9 16 25 36 49 64 81 100 cm

4. Write a super report, modeling your answer on what you learned in the previous 3 activities.

© 1990 by TOPS Learning Systems　　　　32

Answers / Notes

1. The force of gravity (the marble's weight), is resolved into 2 components running parallel and perpendicular to the meter stick. Only the relatively smaller parallel force accelerates the marble down the incline. The perpendicular vector is balanced by the meter stick pushing up in an equal and opposite direction.

2. The clicks sound with increasing frequency as the marble accelerates down the incline.

3. The clicks sound with uniform frequency as the marble accelerates down the incline.

4. SAMPLE REPORT:

The marble accelerates exactly like a floor tapper in free fall, only much slower. Marking time in "clicks" (a much larger units of time than "taps") here is a summary of distance, speed and acceleration as the marble rolls down the incline. Time vs. distance, and time vs. speed are both graphed.

interval (cm)	elapsed time (clicks)	total distance (cm)	avg. speed last click (cm/click)	acc. over previous tap (cm/tap^2)
0-1	1	1	1	————
0-4	2	4	3	2
0-9	3	9	5	2
0-16	4	16	7	2
0-25	5	25	9	2
0-36	6	36	11	2
0-49	7	49	13	2
0-64	8	64	15	2
0-81	9	81	17	2
0-100	10	100	19	2

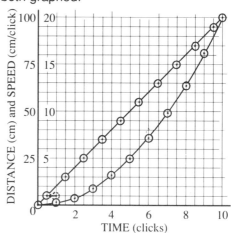

Materials

☐ A meter stick with string guard rails, and a book to raise one end.
☐ A marble.
☐ String.
☐ Scissors.

(TO) clock the speed of a marble through units of distance that increase as perfect squares. To observe that these distances are proportional to time squared.

CLICK ALONG (2)　　　　O　　　　　　Motion (   )

1. Set up a meter stick with string guard rails. Mark the following intervals with a line drawn on masking tape.

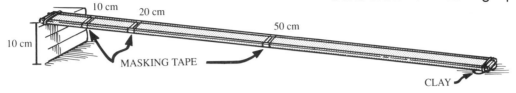

2. Use a stopwatch to time how long it takes a marble to roll across each interval. Use clothespins to define the start and finish lines. Make a data table.

interval (cm)	distance (cm)	time trials (sec) 1	2	3	4	accepted value
10-20	10					
10-50	40					
10-100	90					

3. The distance (d) a marble rolls down the incline, and the square of the time it takes (t^2), are directly proportional. Show that this is true for your data, within the limits of experimental error.

$$d \propto t^2$$

© 1990 by TOPS Learning Systems　　33

Introduction

How can we show that numbers in this data table are directly proportional?

x	y
1	2
2	4
3	6
4	8

- They form equal fractions:
  $1/2 = 2/4 = 3/6 = 4/8$.
  Also, $2/1 = 4/2 = 6/3 = 8/4$.

- They are related by the same constant:
  $1 = .5(2), 2 = .5(4), 3 = .5(6), 4 = .5(8)$, where k = .5.
  Also, $2(1) = 2, 2(2) = 4, 2(3) = 6, 2(4) = 8$, where k = 2.

- They graph into a straight line. (This will be done in the next activity.)

Answers / Notes

1-2. If the marble sometimes delays before starting (after you flip the half clothespin out of the way), raise the ramp a little more. Usually, a 10 cm book platform permits uniform starting, while minimizing the marble's acceleration for easier timing. Some practice with a stopwatch may be necessary before students achieve reasonably consistent time trials.

interval (cm)	distance (cm)	time trials (sec) 1	2	3	4	accepted value
10-20	10	.58	.60	.59	.58	.59
10-50	40	1.17	1.09	1.13	1.19	1.15
10-100	90	1.73	1.70	1.67	1.71	1.70

3. Students should choose one of the methods demonstrated in your introduction.

$$\frac{10}{(.59)^2} = \frac{40}{(1.15)^2} = \frac{90}{(1.70)^2} = k \approx 30 \qquad 10 \approx 30\,(.59)^2, \quad 40 \approx 30\,(1.15)^2, \quad 90 \approx 30\,(1.70)^2$$

Materials

- □ A meter stick with string guard rails.
- □ Masking tape.
- □ A lump of clay.
- □ Books.
- □ A stopwatch.
- □ A marble.
- □ Clothespins.
- □ A calculator (optional).

(TO) graph the distance that a marble rolls down an incline as a function of time squared.

CLICK ALONG (3)　　　　　○　　　　　**Motion (　)**

1. Complete this data table based on your accepted time intervals from the previous activity.

interval (cm)	accepted time (sec)	time² (sec²)	total distance (cm)
10-20			10
10-50			40
10-100			90

2. Graph these relationships on the same graph. Label each graph line.

   a. Time vs. Distance
   b. Time² vs. Distance

3. Interpret your graph lines.

4. If you tilted the incline steeper and steeper, you would finally achieve vertical free fall. How would this affect your graph lines?

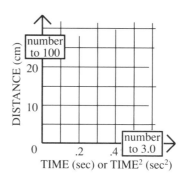

34

Answers / Notes

1.

interval (cm)	accepted time (sec)	time² (sec²)	total distance (cm)
10-20	.59	.35	10
10-50	1.15	1.32	40
10-100	1.70	2.89	90

2.

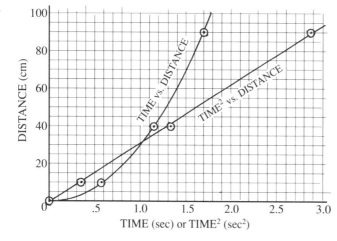

3. The upward-sweeping curve of time vs. distance indicates that the marble accelerates; that it rolls over greater distances with each new unit of time. The straight line graph of time² vs. distance shows that this distance is directly proportional to the square of the time it takes.

4. The acceleration of gravity on the marble would dramatically increase, causing it to travel much greater distances over each new second. Otherwise the shape of each graph line would remain the same. The graph line of time vs distance would still curve up as acceleration increased; the graph line of time² vs distance would still plot as a straight-lined direct proportion.

Materials

☐ Data from the previous activity.

(TO) recognize that gravity accelerates all objects, regardless of weight or mass, by an equal amount.

ROLLING PENNIES 〇 Motion (   )

1. Set up a meter stick incline, 10 cm high, with string guard rails. Raise the strings by sliding clothespin halves under each end.

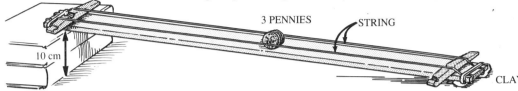

2. Stick 3 pennies together with small pieces of masking tape rolled sticky side out. No tape should stick out beyond the edges.
3. Use a stopwatch to time how long it takes this penny "wheel" to roll down the incline. Complete *only* the top row of this table:
4. As you add more pennies to the roll, *predict* if the rolling-time will increase, decrease or stay the same. Defend your answer by referring to previous experiments.
5. Complete the rest of the table and evaluate your hypothesis. Identify experimental error in this experiment.

| number of | time (sec) | | | accepted |
pennies	1	2	3	value
3				
4				
5				
6				
7				
8				

35

Answers / Notes

4. Adding more pennies to the roll should not affect its total elapsed time down the ramp. Newton's second law predicts that acceleration increases in direct proportion to force and inverse proportion to mass ($a = F/m$). Each new penny increases the weight of the roll, and therefore its force. But this is exactly compensated by a proportional increase in mass. The net effect of increased weight divided by increased mass yields a constant result — the same acceleration.

3, 5. As predicted, there is no change in acceleration as more pennies are added to the roll. Variations that do exist are random, the result of experimental error. A major source of error is the slowing of pennies as they bump into the string guard rails. Random timing error in clicking the stopwatch on and off at the exact instant of starting and ending is also a factor.

More massive objects generally accelerate slightly faster than less massive ones, because their increased inertia enables them to better overcome air resistance and/or surface friction. (Duplicating Galileo's famous experiment of dropping light and heavy weights from a very high place almost works for this reason.) In this experiment however, the extra inertia of a larger penny roll causes it to lean into the string guard rails for longer periods of time, thereby creating more friction. In this "free-fall" system, the extra inertia of heavy-weights gives them no advantage.

| number of | time (sec) | | | accepted |
pennies	1	2	3	value
3	2.00	1.93	1.89	1.94
4	1.83	1.85	1.82	1.83
5	1.91	1.93	1.90	1.91
6	1.96	2.10	2.03	2.03
7	2.07	2.02	1.97	2.02
8	1.90	2.02	1.97	1.93

Materials

☐ A meter stick with string guard rails.
☐ Clothespin halves.
☐ Books.
☐ A lump of clay.
☐ Pennies.
☐ Masking tape.
☐ A stopwatch.

(TO) measure reaction time by catching a dropping meter stick.

REACTION TIME ○ Motion ( )

1. A meter stick falls (d) centimeters in (t) seconds according to the equation: $d = 490\ t^2$. Calculate how far it falls during each time interval listed in the table.

2. Graph your calculations on these coordinates.

time (sec)	distance (cm)
0	0
0.05	
0.10	
0.15	
0.20	
0.25	
0.30	

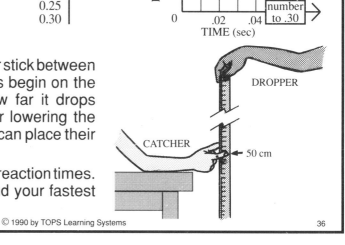

3. Ask a friend to catch a meter stick between thumb and forefinger. Always begin on the 50 cm line, then record how far it drops before being caught. (No fair lowering the hand. To avoid this, catchers can place their arm on a table!)

4. Use your graph to estimate reaction times. Report your average time and your fastest time.

© 1990 by TOPS Learning Systems

36

Answers / Notes

1-2,4.

time (sec)	distance (cm)
0	0
0.05	1.2
0.10	4.9
0.15	11.0
0.20	19.6
0.25	30.6
0.30	44.1

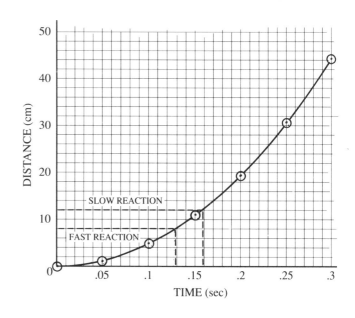

3. *The dropper should give no visual cues that enable the catcher to anticipate the moment of release. Slowly easing your grip on the extreme end of the stick will allows it to fall without warning.*

The catcher should announce "ready" before each time trial begins. The best place to watch the meter stick is at the top.

4. The meter stick typically drops through a range of distances from 8 cm to 12 cm. This corresponds to a range of reaction times between .13 sec and .16 sec.

Materials

☐ A meter stick.
☐ A calculator (optional).
☐ Graph paper.

REPRODUCIBLE
STUDENT
TASK CARDS

Task Cards Options

Here are 3 management options to consider before you photocopy:

1. Consumable Worksheets: Copy 1 complete set of task card pages. Cut out each card and fix it to a separate sheet of boldly lined paper. Duplicate a class set of each worksheet master you have made, 1 per student. Direct students to follow the task card instructions at the top of each page, then respond to questions in the lined space underneath.

2. Nonconsumable Reference Booklets: Copy and collate the 2-up task card pages in sequence. Make perhaps half as many sets as the students who will use them. Staple each set in the upper left corner, both front and back to prevent the outside pages from working loose. Tell students that these task card booklets are for reference only. They should use them as they would any textbook, responding to questions on their own papers, returning them unmarked and in good shape at the end of the module.

3. Nonconsumable Task Cards: Copy several sets of task card pages. Laminate them, if you wish, for extra durability, then cut out each card to display in your room. You might pin cards to bulletin boards; or punch out the holes and hang them from wall hooks (you can fashion hooks from paper clips and tape these to the wall); or fix cards to cereal boxes with paper fasteners, 4 to a box; or keep cards on designated reference tables. The important thing is to provide enough task card reference points about your classroom to avoid a jam of too many students at any one location. Two or 3 task card sets should accommodate everyone, since different students will use different cards at different times.

BODIES AT REST O Motion ( )

1. Run 2 string guard rails down the length of a meter stick. Fix them tightly at each end with rubber bands.

2. Raise one end with 3 books. Mark the other end with masking tape.

3. Cut a "doorway" in a small drinking cup so a marble can roll down the ramp and land inside.

4. Cut a pointer from masking tape to mark how far the cup slides.

5. Repeat with 1 penny, 2 pennies, and 3 pennies taped to the top of the cup.

6. Measure distances. Organize a data table and draw a bar graph on lined paper.

7. Describe how increasing mass (more pennies), affects a body at rest (the cup).

© 1990 by TOPS Learning Systems   1

BODIES IN MOTION O Motion ( )

1. Lower the incline from 3 books to 1 book. Line up the other end with a masking tape baseline.

2. Mark how far the cup slides when one marble rolls off the ramp.

3. Add more marbles. Repeat the experiment after each addition.

4. Measure distances. Organize a data table and draw a bar graph on lined paper.

5. Describe how increasing mass (more marbles), affects bodies in motion.

© 1990 by TOPS Learning Systems   2

SLIDE SHOW (1)   O   Motion ( )

1. Stick 3 pennies and a small button together using small pieces of masking tape rolled sticky side out. (No tape should stick out past any edge on this "slider.")

PENNIES

BUTTON

2. Slide adding-machine tape under string "guard-rails" attached to a meter stick. Slope the meter stick so the slider moves *very slowly* from top to bottom without stopping.

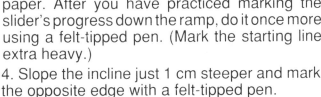

SLIDES SLOWLY

ADDING MACHINE TAPE

3. While a friend calls out "tick-tock" for each passing second, mark the position of your slider (in pencil) along the edge of the paper. After you have practiced marking the slider's progress down the ramp, do it once more using a felt-tipped pen. (Mark the starting line extra heavy.)

24.7 cm PENDULUM   TICK-TOCK

4. Slope the incline just 1 cm steeper and mark the opposite edge with a felt-tipped pen.

   a. Which set of marks describe the fastest moving slider? Explain.

   b. What do the marks say about the movement of your slider?

3

SLIDE SHOW (2)   O   Motion ( )

1. Measure the distance from your starting mark to each succeeding mark. Do this for both the slow and fast tracks.

2. Make a data table and plot your results. Use thread to help you determine the best possible straight line to draw among your circled points.

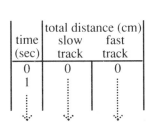

time (sec)	total distance (cm) slow track	fast track
0	0	0
1		

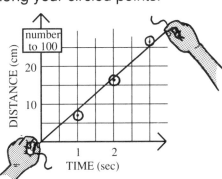

number to 100

3. How is the slope of each graph line related to the speed of your slider?

4. Why don't all points touch the graph line?

4

HEEL-TOE SHUFFLE　　　　O　　　　　Motion (　)

1. Cut a piece of string 5 meters long and lay it on the floor…

…Time how fast you can travel the length of this string, keeping *both* feet on the floor at *all* times.

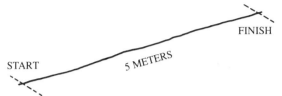

START　　　5 METERS　　　FINISH

YES　　YES　　YES　　NO

2. Construct a data table. Calculate your average speed over 3 trials to the nearest .01 m/sec.

trial	distance (m)	time to shuffle 5 meters (sec)	average speed (m/sec)
1	5		
2	5		
3	5		

$$\text{AVERAGE SPEED} = \frac{\text{TOTAL DISTANCE}}{\text{TOTAL TIME}}$$

3. Does your nose move at a *constant* speed when you heel-toe shuffle? Explain the difference between *average* speed and *constant* speed.

5

HEEL-TOE WALK　　　　O　　　　　Motion (　)

1. Practice walking at a constant speed: touch heel to toe while counting in your head at a calm, steady rate.

1001, 1002, 1003…

UNIFORM WALK

2. Ask a friend to time how long it takes you to walk the same 5-meter string course you used before. Repeat until you get consistent results. Report this as your natural speed.

trial	distance (m)	time to walk 5 meters (sec)	speed (m/sec)
1	5		
2	5		
3	5		

natural speed = ?

3. Does your nose move at a constant speed when you heel-toe walk? Compare this motion to your previous heel-toe shuffle.

4. Using your natural heel-toe walking speed, calculate how far you will travel in exactly 1 minute.

5. Test your prediction in a hallway or outside. Measure distance with your 5 meter string and a meter stick.

6. Evaluate your prediction.

6

HEEL-TOE GRAPH ○   Motion ( )

1. Complete this data table based on your natural heel-toe walking speed from the previous activity.

time (sec)	distance (m)
0	0
10	
20	
30	
60	

2. Graph your results. Label the line "uniform walk."

3. Calculate how far you would travel in 20 seconds if your moved at your fastest average shuffle in activity 5. Plot this value on your graph. Draw a straight dashed graph-line through this point and (0,0), labeling it "fastest average shuffle."

4. Recall that you don't always shuffle at this fastest average speed, that you would soon grow tired. Draw and label a third graph line to show your "predicted progress".

5. What does a graph line's shape and steepness tell you about motion?

7

TRACKING COLLISIONS (1) ○   Motion ( )

1. Fold an index card into a long "M"…

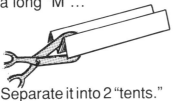

Separate it into 2 "tents."

2. Fold up a 3 cm flap on each tent.

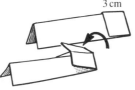

3. Pinch each tent closed while you force open the flap.

PINCH

4. Turn over the tents to make ramps. Practice rolling a marble down each one so they collide on scratch paper in between.

5. Dip each marble in soapy colored water. Record these collisions on paper…

(a) Head-on: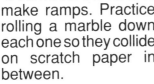

(b) Glancing blow:

Mark the tracks with arrows. Draw in the outline of each marble at the point of impact.

6. What could you conclude about the mass of your marbles if you saw only the path they left and nothing more? (This is similar to tracking subatomic particles in a cloud chamber.) *Save your ramps.*

8

TRACKING COLLISIONS (2) ◯   Motion (   )

1. Record these collisions between a marble and a ping pong ball, using colored soapy water on paper :
    (a) Head on;
    (b) Ping Pong ball strikes resting marble;
    (c) Marble strikes resting Ping Pong ball.

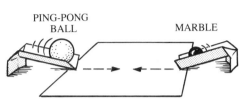

PING-PONG BALL   MARBLE

2. Mark the tracks with arrows. Draw in the outline of each sphere at the point of impact.

3. Which ball most resisted changing...
    a. Its state of motion? Explain.
    b. Its state of rest? Explain.

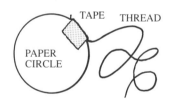

4. Roll some clay between your palms until it is the size and shape of a glass marble. Drop both onto your table top.

CLAY   MARBLE

    a. Describe the *elastic* collision between the *glass* marble and the table.
    b. Describe the *inelastic* collision between the *clay* marble and the table. (Save your ramps.)

9

OUT FROM UNDER ◯   Motion (   )

1. Trace around the mouth of a baby food jar on an index card. Cut out the circle and tape its edge to some thread about as long as notebook paper.

TAPE   THREAD

PAPER CIRCLE

2. In each case, remove the circle, but leave the penny in place:

(a) EASY:
penny on a clothespin.

(b) HARDER:
penny on your finger.

(c) HARDEST:
penny on your finger; no thread on the circle.

3. Explain these tricks in terms of Newton's first law of inertia.

10

BOOK DROP? 〇   **Motion ( )**

1. Wrap string several times around a heavy book and tie it. Loop thread through this string on both sides.

2. Wrap one thread loop securely around your pencil. Tape the free end.

3. Suspend the book by holding the pencil along your table edge. Put a coat or something soft underneath.

4. Pull the bottom loop as directed below. Explain what happens in words and pictures:

    a. *Rapidly* pull the bottom loop with a hard, fast jerk. Does the book drop?
    b. Repeat the experiment. This time *slowly* pull the bottom loop. Does the book drop?

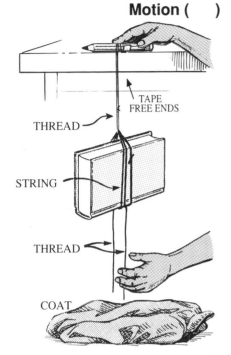

11

THE HOOP TRICK 〇   **Motion ( )**

1. Cut off 4 strips of index card exactly 3 spaces wide. Overlap them (in the same direction) about the width of your little finger, and secure with tape.

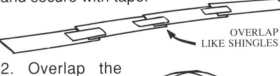

OVERLAP LIKE SHINGLES

2. Overlap the ends, keeping the tape to the outside of the circle. Tape again to form a closed hoop.

NO INSIDE TAPE NEEDED

3. Balance the hoop over a bottle or Erlynmeyer flask. Crunch together a small wad of aluminum foil to balance on top. (It should easily fit through the mouth of the bottle.)

4. Drop the foil into the bottle by touching only the hoop with only *one* finger. Learn this trick, then challenge your friends.

5. Explain why this trick works. Illustrate your answer with a diagram.

12

TENSION Motion ( )

1. Stretch 12 rubber bands of uniform size to their limit, then string them between 13 unbent paper clips. Stretch this chain across your floor between 2 rubber-banded textbooks so each band stretches to about 3/4 capacity.

12 BANDS
13 CLIPS

    a. Are all forces in this system balanced? How do you know?
    b. Diagram and label the *horizontal* forces that act on 1 of the paper clips.
    c. Diagram and label *all* forces acting on 1 of the books.
    d. Does each rubber band have the same tension as it neighbor? Explain.

2. Without moving the books, measure the length of all 12 stretched bands (from left to right), to the nearest .1 cm. Make a data table.

band number	length (cm)
1	
2	
3	
⋮	

3. Select 4 rubber bands with nearly the same length (within 1 cm). Fix each to its paper clip with masking tape, write its length on the tape, and save.

13

MAKE A SCALE Motion ( )

1. Cut a strip of cardboard about 2 cm wide and 30 cm long. Draw a line down the middle, crossing it 1 cm from an end. Paper-punch a hole just under this intersection.

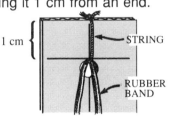

1 cm

STRING

RUBBER BAND

2. Choose a tagged rubber band from the previous activity. Tie it snugly *inside* the punched hole with string.

3. Suspend a spring balance from your table edge with a loop of string fixed with masking tape. Hang the cardboard underneath, through its hole and through the rubber band.

4. Mark where the bottom tip of the rubber touches the cardboard. Call this "0."

5. Pull *only* the paper clip to the next major division on the spring scale, letting the cardboard hang free. Pinch both together at that point and mark again where the bottom of the rubber band touches. Calibrate in this manner all the way down.

6. Describe how your rubber band responds to equal amounts of force. (Save your scale.)

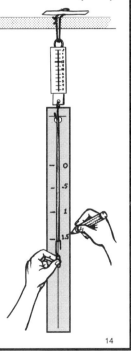

14

TWO KINDS OF FRICTION Motion ( )

1. Tie a chain of several rubber bands to an empty milk jug. Hook your rubber band scale on the other end.

2. Fill the jug with enough water (perhaps 1/4 full) so you can move it just before exceeding the capacity of your scale.

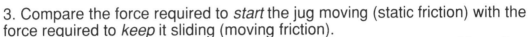

3. Compare the force required to *start* the jug moving (static friction) with the force required to *keep* it sliding (moving friction).

   a. Are these 2 kinds of friction overcome by the same amount of force? Support your answer with numbers.

   b. Propose a hypothesis to explain your observations.

4. As you pull the scale with more and more force, rubber bands on the jug pull in an equal and opposite direction against you!

   a. What evidence can you observe to support this idea?

   b. Why does the jug eventually move? Write about balanced and unbalanced forces as you answer.

15

BALANCED FORCES  Motion ( )

1. Hook your remaining 3 nearly-equal rubber bands into a central washer. Stretch them out using 3 heavy rubber-banded books as ballast.

WASHER

2. Measure each unknown angle with a protractor, and each unknown force with your scale calibrated in Newtons. (Tape the rubber band and paper clip out of the way on the back of the cardboard.) Write about your discoveries.

(a) Three equal forces.	(b) Two against one.	(c) Bending a straight line.	(d) pulling at a right angle.
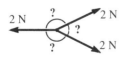	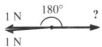	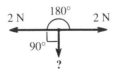	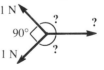

16

BUILD AN ACCELEROMETER ○   Motion ( )

1. Fix a paper clip and thread with masking tape under the lid of a baby food jar, so it hangs in the center when closed, not quite touching the bottom.

2. Fill the jar with water and tightly close the lid.

3. Push your accelerometer so it slides untouched across a table and comes to rest. (Don't let it fall off the edge!) Explain in words and pictures how the paper clip shows acceleration (speeding up); deceleration (slowing down).

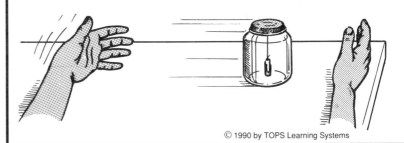

4. Walk about the room holding your accelerometer. What can you discover?

© 1990 by TOPS Learning Systems   17

ACCELERATION IS CHANGE! ○   Motion ( )

ACCELEROMETER

In each case below…
   a. Identify the acceleration acting on your accelerometer (if any), naming the balanced or unbalanced forces involved.
   b. Draw a diagram showing force vectors.

1. Push the jar to *start it moving.*

2. Allow a moving jar to *come to a free sliding rest.*

3. Move the jar across your table *in a straight line, at constant speed.* (Maintain continuous contact with your hand.)

4. Observe the jar *at rest.*

5. Move the jar on your table *in a circle at constant speed.* (Diagram this with a top view.)

© 1990 by TOPS Learning Systems   18

SECONDS PER SECOND ○ Motion (   )

1. Each square below represents 1 meter that you travel on a bicycle. Fully describe your motion (in numbers plus units) from left (t = 0 sec) to right. The first "voyage" is done as an example.

a. →[ ]→ -1 0 1 2 3 4 5 6 7   8   9   10 SECONDS

(You are moving at a constant speed of 1 m/sec. At t = 7 sec, you begin to accelerate 2 m/sec each new second, or 2 m/sec².)

b. st[ ]→   0   1   2   3   4

c. st[ ]end   0 1   2   3   4   5   6   7   8 9

d. →[ ]→   -1   0   1   2   3   4   5   6,7,8,9 / 10   11   12   13

2. Graph and label each "voyage" on the same pair of coordinates, beginning at (0,0). Check your descriptions against each graph line.

DIST (m) / TIME (sec)

19

TRACKING A CURVE ○ Motion (   )

1. Lean a manila folder on a book. Bridge the gap at the bottom with scratch paper.

2. Slide your index card ramp up to the corner of this incline so a marble rolls up and back down, traveling across most of the paper. (Set the ramp on a lump of clay if necessary.)

3. Dip a marble in soapy colored water and let it trace a path across your paper.

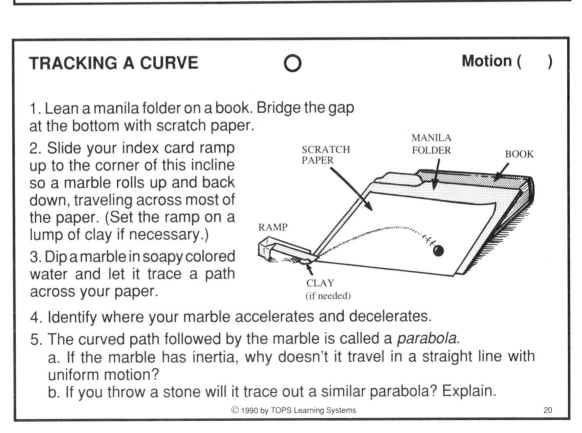

SCRATCH PAPER
MANILA FOLDER
BOOK
RAMP
CLAY
(if needed)

4. Identify where your marble accelerates and decelerates.

5. The curved path followed by the marble is called a *parabola*.
   a. If the marble has inertia, why doesn't it travel in a straight line with uniform motion?
   b. If you throw a stone will it trace out a similar parabola? Explain.

20

EARTH-MOON MODEL ⭕ Motion ( )

1. Tie a paper clip to both ends of a 30 cm thread that has been pulled through a 10 cm straw. Add 5 additional paper clips to one end.

2. Spin the single paper clip just fast enough to support the other 6 clips about 1 cm from the bottom of the straw.

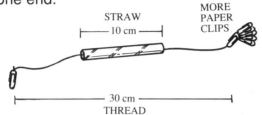

STRAW
|— 10 cm —|
MORE PAPER CLIPS

|———— 30 cm ————|
THREAD

1 cm

3. Let this system model the earth and moon.

    a. What represents the moon? the earth? gravity?
    b. Why doesn't the moon come crashing to earth?
    c. If the moon's orbit slowed, would it be pulled straight to earth? Explain.
    d. Suppose Earth's moon had twice the mass. Would it circle the earth at the same rate? Back up your answer with experimental evidence.

21

FORCE OVER MASS... ⭕ Motion ( )

1. Push your accelerometer using different amounts of unbalanced force. Summarize your results in a table.

2. Newton recognized that acceleration is *directly proportional* to force. How do your results support his observation?

unbalanced force	acceleration
small push	
medium push	
large push	

3. Fill 2 more baby food jars with water and close the lids. Practice applying equal force, first to 1 jar, then 2, then 3 in a row. Do this by pushing as hard as you can, using only the *back* of your little finger, and keeping your wrist on the table.

WRIST DOWN

4. Summarize your results in a table. How do your results support Newton's observation that acceleration is *inversely proportional* to mass?

mass	acceleration
small (1 jar)	
medium (2 jars)	
large (3 jars)	

5. Newton's second law is summarized by this equation: What does it mean?

$$a = \frac{F}{m}$$

22

FREE FALL **Motion (   )**

1. Make 3 penny weights like these. Hold them together with masking tape.

1 PENNY   2 PENNIES   4 PENNIES

2. Gravity pushes down on 1 penny with a force of .025 Newtons. (This is its *weight*.) Calculate the weight of 2 pennies; of 4 pennies.

3. One penny has a *mass* of .0025 kg. Calculate the mass of 2 pennies; of 4 pennies.

4. In free fall, an object is pulled down by the force of its own weight. Calculate how fast each mass accelerates in free fall.

$$a = \frac{F}{m}$$

5. People commonly think that heavy objects fall faster than light objects.
   a. What do your results in step (4) say about this?
   b. Drop all 3 penny weights together and write what you observe.

23

AIR RESISTANCE **Motion (   )**

1. Divide a sheet of scratch paper exactly in half. Crumple one, leave the other flat.

2. Crumple the flat sheet around 10 pennies until it is the same size as the other.

10 PENNIES INSIDE

a. Compare their rate of free-fall.
b. What is the real cause of objects not falling together? Draw a labeled diagram to show the forces acting on each paper.

a. Now that they have the same air resistance, do they free-fall together?
b. What role does mass play in overcoming air resistance? Explain.

3. Put a penny and a wisp of cotton (or a feather) in a paper cup.
   a. Predict what will happen when you drop the cup. Refer to Newton's second law.   b. Test your prediction.

24

CLOTHESPIN LAUNCHER Motion (   )

1. Fix a clothespin to a baby food jar with two rubber bands so the jaws point up.

2. Wedge the clothespin wide open with another half clothespin tied to a trip thread.

3. Put this next to a baby food jar lid in an open area on the floor. Pull the thread with a *quick* flick of your wrist. Describe what happens.

4. Interpret your observations in terms of Newton's third law.

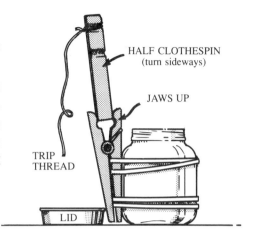

HALF CLOTHESPIN
(turn sideways)

JAWS UP

TRIP
THREAD

LID

5. According to Newton's second law, the acceleration of any object is inversely proportional to its mass. Demonstrate this by pressing a lump of clay into the lid. What can you conclude?

25

JET STRAW Motion (   )

1. Attach a balloon to the end of a flexible straw with masking tape.

2. Push a straight pin through the straw near the balloon. Pivot this on another vertical straw.

3. Blow up the balloon, then release it and watch it spin.

   a. How does the angle of the bend affect the performance of your jet straw?

   b. Explain your observations in terms of Newton's third law.

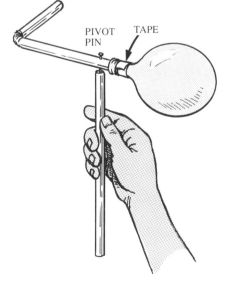

PIVOT
PIN

TAPE

26

CATAPULT (1) ○ Motion (   )

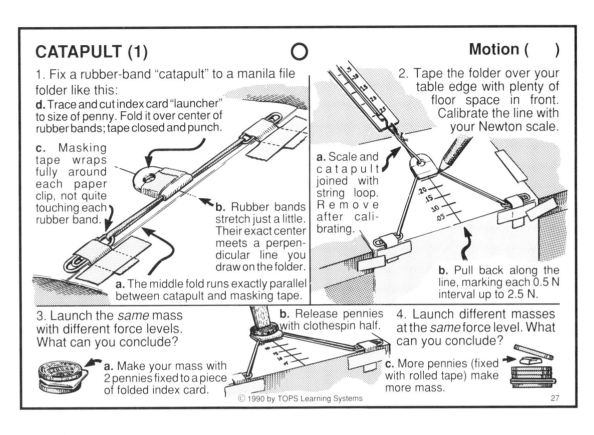

1. Fix a rubber-band "catapult" to a manila file folder like this:

d. Trace and cut index card "launcher" to size of penny. Fold it over center of rubber bands; tape closed and punch.

c. Masking tape wraps fully around each paper clip, not quite touching each rubber band.

b. Rubber bands stretch just a little. Their exact center meets a perpendicular line you draw on the folder.

a. The middle fold runs exactly parallel between catapult and masking tape.

2. Tape the folder over your table edge with plenty of floor space in front. Calibrate the line with your Newton scale.

a. Scale and catapult joined with string loop. Remove after calibrating.

b. Pull back along the line, marking each 0.5 N interval up to 2.5 N.

3. Launch the *same* mass with different force levels. What can you conclude?

a. Make your mass with 2 pennies fixed to a piece of folded index card.

b. Release pennies with clothespin half.

4. Launch different masses at the *same* force level. What can you conclude?

c. More pennies (fixed with rolled tape) make more mass.

© 1990 by TOPS Learning Systems   27

CATAPULT (2) ○ Motion (   )

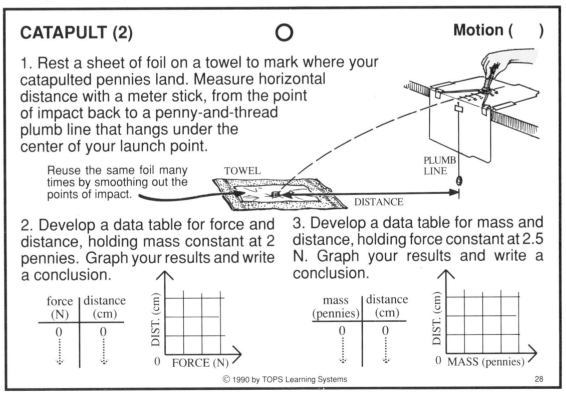

1. Rest a sheet of foil on a towel to mark where your catapulted pennies land. Measure horizontal distance with a meter stick, from the point of impact back to a penny-and-thread plumb line that hangs under the center of your launch point.

Reuse the same foil many times by smoothing out the points of impact.

TOWEL

PLUMB LINE

DISTANCE

2. Develop a data table for force and distance, holding mass constant at 2 pennies. Graph your results and write a conclusion.

force (N)	distance (cm)
0	0
⋮	⋮

DIST. (cm)

0 FORCE (N)

3. Develop a data table for mass and distance, holding force constant at 2.5 N. Graph your results and write a conclusion.

mass (pennies)	distance (cm)
0	0
⋮	⋮

DIST. (cm)

0 MASS (pennies)

© 1990 by TOPS Learning Systems   28

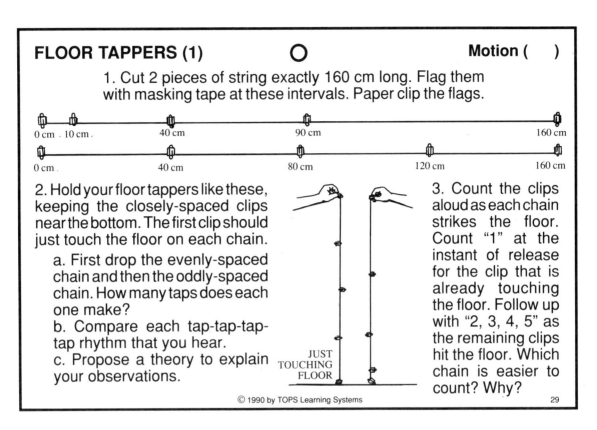

FLOOR TAPPERS (1)   O   Motion (   )

1. Cut 2 pieces of string exactly 160 cm long. Flag them with masking tape at these intervals. Paper clip the flags.

0 cm . 10 cm .   40 cm   90 cm   160 cm

0 cm .   40 cm   80 cm   120 cm   160 cm

2. Hold your floor tappers like these, keeping the closely-spaced clips near the bottom. The first clip should just touch the floor on each chain.

a. First drop the evenly-spaced chain and then the oddly-spaced chain. How many taps does each one make?
b. Compare each tap-tap-tap-tap rhythm that you hear.
c. Propose a theory to explain your observations.

JUST TOUCHING FLOOR

3. Count the clips aloud as each chain strikes the floor. Count "1" at the instant of release for the clip that is already touching the floor. Follow up with "2, 3, 4, 5" as the remaining clips hit the floor. Which chain is easier to count? Why?

© 1990 by TOPS Learning Systems   29

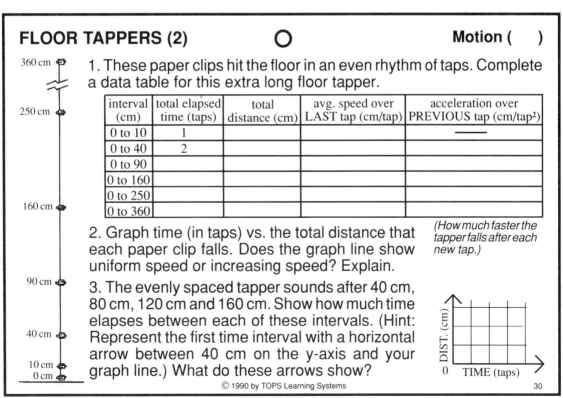

FLOOR TAPPERS (2)   O   Motion (   )

360 cm

250 cm

1. These paper clips hit the floor in an even rhythm of taps. Complete a data table for this extra long floor tapper.

interval (cm)	total elapsed time (taps)	total distance (cm)	avg. speed over LAST tap (cm/tap)	acceleration over PREVIOUS tap (cm/tap²)
0 to 10	1			——
0 to 40	2			
0 to 90				
0 to 160				
0 to 250				
0 to 360				

160 cm

2. Graph time (in taps) vs. the total distance that each paper clip falls. Does the graph line show uniform speed or increasing speed? Explain.

(How much faster the tapper falls after each new tap.)

90 cm

40 cm

10 cm
0 cm

3. The evenly spaced tapper sounds after 40 cm, 80 cm, 120 cm and 160 cm. Show how much time elapses between each of these intervals. (Hint: Represent the first time interval with a horizontal arrow between 40 cm on the y-axis and your graph line.) What do these arrows show?

DIST. (cm)

0   TIME (taps)

© 1990 by TOPS Learning Systems   30

FLOOR TAPPERS (3)   〇   Motion (   )

1. Graph the average *speed* of each paper clip over its last interval, just before it hits the floor. (Use the 2nd and 4th columns from the data table in the previous activity.)

Note: Shift each point 2 squares (1/2 tap) left, since the average speed happens half-way through each interval.

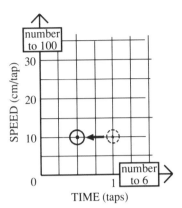

2. Does the graph line show uniform acceleration? Explain.

3. Draw vertical arrows on your graph to show the increase in speed (acceleration) over successive time intervals. Compare your answer with the previous data table.

31

CLICK ALONG (1)   〇   Motion (   )

1. Set up a meter stick with string guard rails. Explain how inclines are able to slow the rate of a free-falling marble.

force that accelerates marble

marble presses into meter stick

gravity pulls straight down

2. Lay strings across the meter stick at regular 10 cm intervals. Listen to a marble as it click, click, clicks down the incline. Report your observations.

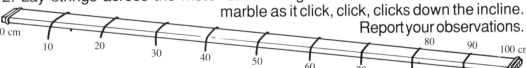

3. Repeat the experiment for strings placed at these perfect-square intervals. Listen again as a marble clicks down the ramp.

4. Write a super report, modeling your answer on what you learned in the previous 3 activities.

32

CLICK ALONG (2)　　　　O　　　　Motion (　)

1. Set up a meter stick with string guard rails. Mark the following intervals with a line drawn on masking tape.

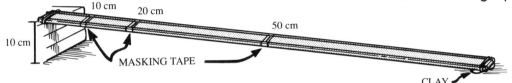

2. Use a stopwatch to time how long it takes a marble to roll across each interval. Use clothespins to define the start and finish lines. Make a data table.

HALF CLOTHESPIN:
FLIP TO START

START:
10 cm

FINISH: 20 cm
or 50 cm

interval (cm)	distance (cm)	time trials (sec) 1	2	3	4	accepted value
10-20	10					
10-50	40					
10-100	90					

3. The distance (d) a marble rolls down an incline, and the square of the time it takes (t^2), are directly proportional. Show that this is true for your data, within the limits of experimental error.

$$d \; \alpha \; t^2$$

33

CLICK ALONG (3)　　　　O　　　　Motion (　)

1. Complete this data table based on your accepted time intervals from the previous activity.

interval (cm)	accepted time (sec)	time^2 (sec^2)	total distance (cm)
10-20			10
10-50			40
10-100			90

2. Graph these relationships on the same graph. Label each graph line.

   a. Time vs. Distance
   b. Time^2 vs. Distance

3. Interpret your graph lines.

4. If you tilted the incline steeper and steeper, you would finally achieve vertical free fall. How would this affect your graph lines?

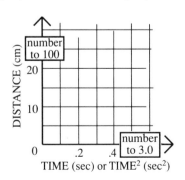

number to 100

number to 3.0

DISTANCE (cm)

TIME (sec) or TIME^2 (sec^2)

34

ROLLING PENNIES   Motion ( )

1. Set up a meter stick incline, 10 cm high, with string guard rails. Raise the strings by sliding clothespin halves under each end.

2. Stick 3 pennies together with small pieces of masking tape rolled sticky side out. No tape should stick out beyond the edges.

3. Use a stopwatch to time how long it takes this penny "wheel" to roll down the incline. Complete *only* the top row of this table:

4. As you add more pennies to the roll, *predict* if the rolling-time will increase, decrease or stay the same. Defend your answer by referring to previous experiments.

5. Complete the rest of the table and evaluate your hypothesis. Identify experimental error in this experiment.

number of pennies	time (sec) 1	2	3	accepted value
3				
4				
5				
6				
7				
8				

35

REACTION TIME ◯ Motion ( )

1. A meter stick falls (d) centimeters in (t) seconds according to the equation: $d = 490 \, t^2$. Calculate how far it falls during each time interval listed in the table.

2. Graph your calculations on these coordinates.

time (sec)	distance (cm)
0	0
0.05	
0.10	
0.15	
0.20	
0.25	
0.30	

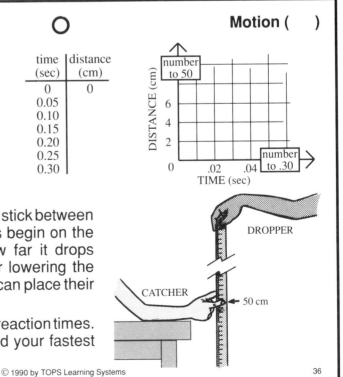

3. Ask a friend to catch a meter stick between thumb and forefinger. Always begin on the 50 cm line, then record how far it drops before being caught. (No fair lowering the hand. To avoid this, catchers can place their arm on a table!)

4. Use your graph to estimate reaction times. Report your average time and your fastest time.

36

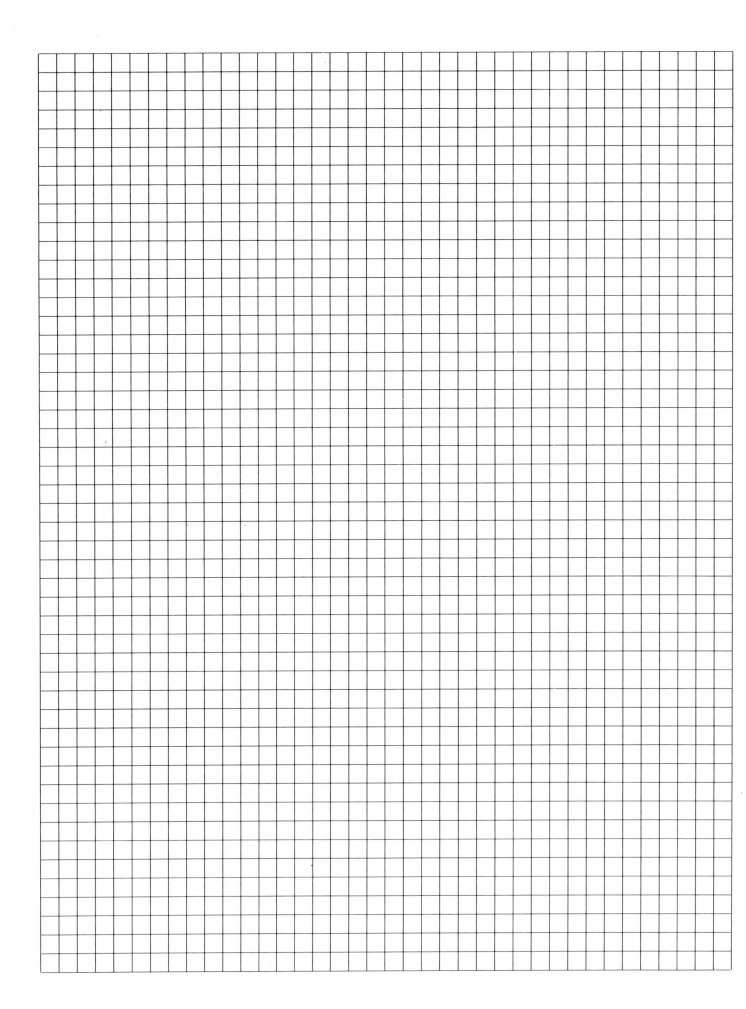